HISTORY OF SCIENCE AND TECHNOLOGY

ELECTRIC POWER

INTRODUCTION

This work examines the issues of historical development tia of science and technology, electric power industry.

This knowledge is important both for students studying in this direction, and for specialists working in electrical energy.

Knowledge of the history of the development of science and technology, this most important government activities of any state, allows you to correctly assess the current situation in the electricity industry whether to take into account the experience of previous generations and develop the industry taking into account these factors.

The development of the electric power industry is a powerful force that affects living standards of people, changes the character of society, is cause social change and guides social development.

The content of this discipline is structured in such a way that the student could first of all immediately cover the range of basic discipline issues, familiarize yourself with keywords and understand tions, their origin and meanings, first in a general form, and topics to study the content of each of the concepts in more detail history, science, technology, technology, progress, engineer, energy,energy, electric power, etc.

For the convenience of studying the discipline in the first part of the work, brief definitions of key words, their basic conceptual meanings and then a separate paragraph is devoted to each of the titles, revealing their essence in more or less detail.

The following sections provide a history of scientific discoveries and inventions in semantic and chronological order, their technical use in the field of electricity and electricity.

Attention is drawn to the prospects for further development of nonwhich areas in the field of generating electricity at no power sources.

The end of the manual briefly covers the impact of work electric power devices for the environment.

Keywords and concepts in the discipline

History (gr. Historia) - research: 1) reality in the course its development, movement; 2) the past, preserved in human memory honor; 3) the science of the development of human society, the field of knowledge, nature, technology, progress, culture, etc.

Science - 1) what teaches, gives experience, lesson; 2) production and system thematization of objective knowledge about reality; 3) system knowledge about the laws in the development of nature, society, thinking, technology, culture, etc.

Progress - (lat.progressus) - 1) moving forward, to a more perfect shenny state; 2) change for the better; 3) transition to more high level of development.

Technique - (gr. Techne) - craft, craftsmanship, art.

Technology - (gr. Techne + logos (teaching)) - a body of knowledge on the methods and means of carrying out the processes.

Engineer - (lat.ingium - invention, ingenuity; fr. ingenieur) - a sharp inventive mind.

Energy - (gr. Energeia - activity) - 1) one of the main properties of matter - the measure of its movement; 2) active power; 3) general quantitative measure of various forms of motion of matter (field).

Energy - methods and means of application and operation of personal types of energy for industrial, transport, agricultural economic and other needs.

Electricity is a leading energy sector - complex technical education closely interacting with fuel business and the main industries of the mining and processing industry, transport and agriculture.

1. HISTORY. THE SCIENCE. TECHNICS. ENERGY

1.1. History

History is a study, a set of facts, events, related going to the past life of mankind, some branch of science or technology, object, culture, etc. This is a memory of the past, of the events, people. The concepts of "history", "historical fact" include not only past events, but also what has attitude to a person, to his inner world. History brings up and forms a person, she is a great teacher of a person and society.

The purpose of each generation of people is to:
- to master what was left by past generations;
- bring new things with your labor - create, build, research,improve;
- to pass it on to the future generation.

By communicating knowledge in the form in which it was first obtained, history shows the methods of work and the course of creative thought, teaches courage and initiative, fosters a sense of the new and encourages action. It is correct to say that history is the science of the future. If we want to manage our future and guide it to the right goals, then one of the main means for this is the study of the history rii - to reveal how developed and develops, for example, our society, or science and technology, and how to use in our time the most useful history lessons.

The study of the historical past is of practical importance, since its final conclusions bring us close to practical the needs of the moment.

The history is inexhaustible and endless. It continues now. We are contemporaries and witnesses of many inventions, discoveries that are and will make many changes in the mother the physical, social, cultural and spiritual life of mankind, in world knowledge is radio, television, nuclear and thermonuclear energy, computer technology, nanotechnology, etc.

Development of electrical engineering and power engineering - ours with you areas of research - associated with the work of so many people: hescientists, inventors, inquisitive people, scientists - not indifferent ny, thinking, hardworking, morally rich people.

Studying the history of the electric power industry is a kind of communication, affects the incentive and motivational aspects personality, economic needs. This produces the effect presence, which allows you to be, as it were, an accomplice to consider events to be taken.

Based on the spiritual and moral experience of centuries, penetrate repenting with historical feeling, a person gradually develops in itself personal responsibility for all the past and

walking in the world. A sense of moral duty is strengthened in him, which is the core of the true personality.

The history of science and technology inspires belief in the surmountability of difficult stey, into the limitless possibilities of man.

1.2. The science

Science at each considered moment of time represents the result is a body of knowledge about nature, society, thinking, accumulated in the course of the social and historical life of people.

The purpose of science is to reveal the objective laws of phenomena, to give them explanation.

The tasks of science are to find and explore for random, chaotic objective laws, hidden from a superficial glance, and to live by the knowledge of these laws of people for their practical activities.

There are four stages of cognition in the history of mankind nature.

The first stage begins from ancient times (Archimedes, Thales

Miletsky and others) and ends around the 15th century. It formsXia syncretic , that is, an undetermined view of the the driving world; but already in the XIII-XIV centuries ideas and conjectures were born ki, which became the beginning of the formation of natural sciences.

The second stage - XV-XVI centuries. - called analytical, since during this period thinking begins to focus on the

division of concepts and the allocation of particulars, which led to the emergence the development and development of sciences: astronomy, physics, chemistry, biology, and others.

Third stage - XVII-XX centuries; it is called synthetic. In it time, there is a gradual reconstruction of a holistic picture of childbirth based on previous experience.

The fourth stage is the end of the XX century, the beginning of the XXI century. Here begins to form an integral-differential approach to cognition

nature, that is, a single science of nature is considered. Universe Naya, Life, Reason - are interpreted as a single, but very multifaceted object of natural science.

Forecast of the future - a leading role in further knowledge nature belongs to the synthesis of knowledge, the integration of sciences, in the center of there will be a person.

Science is multifaceted, multifaceted and complex phenomenon. Science is also the experimental means necessary for the study of phenomena - these include instruments and installations, with by which these phenomena are recorded and reproduced; this is and the methods by which objects are distinguished and cognized research; these are people engaged in scientific research; this and knowledge systems fixed with the help of texts , etc.

The general basis of the listed phenomena is the technology human activity for the production of knowledge, that is, science ka is a certain human activity that is aimed at gaining knowledge.

The development of science and technology always occurs in specific sources cultural and cultural conditions, determined primarily by the production the leading forces of society, the mode of production. Achieved science and technological progress at the same time contribute to the evolution of society, generating and determining the level of production strong forces.

At first, science took a lot from the masters - engineers of the era of the
birth, then, in the XIX-XX centuries. engineering activity has become already in accordance with science. Specialization and professional
zation of science and technology led to the technization of science and humanization technology.

The whole history of mankind shows that science develops under the influence of practical needs and, above all, the need stey of production. However, production needs do not determine all the complex dynamics of the formation of knowledge, the creation of new ideas, theories, conclusions. It has its own laws. For example
measures, the history of the development of natural sciences gives many examples of scientific discoveries that were not directly generated by queries life:

- the discovery of electricity,
- diffraction,
- magnetism,
- polarization,
- the periodic table of elements and much more.

Science fiction writers, artists, writers are sometimes capable of invade the unknown future and discern the contours of future things kov. Here are some examples.

Englishman Roger Bacon in the XIII century. (about 1240s) understudied languages, mathematics, astronomy, physics, chemistry and made numerous important discoveries. He wrote about chariots, moving with incredible speed without the help of animals, about flying machines, about the properties of concave and convex glasses for the eyes (glasses), created the telescope theory and much more. One of the greatest minds humanity!

In the XVI century (1560s) Francis Bacon (Roger's namesake) created one of the brilliant creations of the human mind - the product denial "New Atlantis", in which he outlined the draft state military organization of science, described the basics of the logic of the renewal of science, pointed out the possibility of

useful application of the observed in nature de phenomena, predicted the creation of submarines, aircraft, movies, radio, television, bionics, a thermonuclear reactor and many other goe.
In the novel by Russian science fiction writer N. Shelonsky "In the world of the future" (1892) we are talking about the transformations of elements, about synthetic materials riyal, about the transmission of thoughts at a distance, about antigravity and many hom friend.
A. Tolstoy in the work "The Hyperboloid of Engineer Garin" described the laser project in detail.
I. Efremov's story "Shadows of the Past" (1945) prompted the
Y. Denisyuk for the discovery of volumetric holography.
Of the 86 predictions of science fiction writer H. Wells, more than
30, and science fiction writers Jules Verne and A. Belyaev came true 90% of their predictions call.
In connection with the above, we consider it appropriate to cite a wise a popular proverb: "The new is the long forgotten old." It is important remember each generation and show interest and respect for work
I will give to the fantasies of people who lived before us.

Next, we will consider some laws of the development of science.
First law. It is called the law of relative independence.
the development of science.
This relative autonomy includes internal old logic of development , the need for systematization of knowledge , struggle of opinions , mutual influence of sciences, interaction with different forms of public consciousness , the continuity of ideas , etc. - then there are all those factors from which, apart from the need for production management (or household), the development of science depends.
Second law. The following law reflects phenomena such as criticism and struggle of opinions in science. That is, the development of science occurs based on the struggle of new and old ideas. Excluding emotional
discussions of new knowledge with old, without a correct understanding of the waste It is impossible to understand the progress of science as a whole in science.

The history of science is the history of the change of various theories and their struggle would. Incompleteness, imperfection of knowledge inevitably leads to that the same series of observed facts receives different explanations different scientists, they see these facts as if from different angles owls. It depends on the difference in views, mentality, talent and etc. However, over time, science inevitably comes to a unified look at them.

Third law. This law expresses the interaction of sciences and has now it is especially important to understand the ongoing processes of scientific and technological progress.

Science is a single whole. Existing sectionthe spread of science into separate areas is due to the difference in nature things and patterns that these things obey in the process movement and development. Various fields of science are developing, interrelated acting with each other in different ways:

- through the use of knowledge gained by other sciences;
- through the use of methods of studying other sciences;
- through technology and production;
- through the study of the general properties of various types of matter.

The fourth law characterizes the process of mathematization of practice all scientific disciplines. Mathematics is now penetrating even into such areas of knowledge as history, linguistics, biology, etc.

the power of a computer decrypts ancient Mayan manuscripts, etc. many branches of physics, astronomy - mathematics is indispensable silent apparatus.

The fifth law relates to the differentiation and integration of sciences, which are invariably present in the development of modern natural knowledge.

The process of differentiation - the rebirth of various branches sciences into independent scientific disciplines . However, this the process is associated with an integration process linking different branch of natural science , since there is a rapid development of frontier sciences: genetic engineering, molecular geology, biogeochemistry, engineering psychology, etc.

The sixth law is continuity in science. Science represents cofight is a product of the activities of many generations. Its objective content zhania is not eliminated along with the elimination of the social system, but develops and accumulates throughout the history of human state. Use and development of knowledge accumulated by previous

generations, that is, continuity , is an object
tive law of the development of science. Without him, nothing is simply

impossible development!

The seventh law, discovered by F. Engels - accelerated development science is still valid. The achievements of the 19th century were many times over the achievements of the 18th century go back, and the achievements of the 20th century (even the second half of it) surpass the achievements of previous times.

The eighth law testifies to the inevitability of scientific revolution lucius. Analysis of the history of the development of natural science shows that it developed very unevenly. Periods of relative stability

ness, the gradual accumulation of knowledge is inevitable over time times were replaced by shorter periods of revolutions, which when there is a radical breakdown of theoretical concepts, hiding previously unshakable.

The ninth law describes the strengthening of the connection between science and production, which ultimately led to the understanding of science as one of the most important elements of the productive forces. As a result, a technogenic naya civilization, which is replaced by human civilization or postindustrial society.

Science is the creation of life. From the life around scientific man's thought takes, given in the form of scientific truth, term. Science is a manifestation of action in human society ve total human thought.

V.I. Vernadsky.

1.3. Technics

Technique - (gr. - art, craft, craftsmanship). Development of technical niks , including energy and power engineering, - there is a powerful system la , which affects the living standards of people , changes the character society , its social system - primitive, slavesky, feudal, serf, capitalist, social

static, etc. The development of technology is the cause of social changes and guides social development .

Technology at the end of the 19th-20th centuries was in the focus of study various disciplines, both technical and natural, as well public.

Technology can be attributed to the sphere of material culture .

With her adaptations, devices, she strengthened our hearing,

tion, strength and dexterity; it shortens distance and time, increases labor productivity. Making it easier to meet needs, it thereby creates new needs.

Technology gave humanity space and time, matter and force, and itself serves as the force that irrepressibly drives forward forest of progress.

But! Material culture is closely related to a person with spiritual culture is the most inseparable bonds, and therefore technology has and intangible aspect in the form of a body of knowledge .

Science and technology are built from facts and experience. But not organized a simple collection of facts, experiences, observation niy as little deserves the name of science and engineering as a pile bricks deserves the name of the house.

Technology has its own objective laws of structure and development (like science). Due to the inevitable development and complication of the world technology, accelerating scientific and technological progress all knowledge about the technique itself as a whole is becoming more relevant - unified ideas about the structure and development of a wide variety of materials tires, devices, apparatus. Therefore, an extension of general technical technical and fundamental training of engineers.

Research and formulation of objective laws of structure and the development of technology by analogy with the laws of nature - one of main and little developed areas of general technical fundamentalization of engineering education.

Ak. I. F. Obraztsov. The formation of a unified system of scientific knowledge introduces into the field technology the concept of "technosphere" (by analogy with the biosphere, noosphere, atmosphere, etc.). This concept fixes some integral passages parameters of natural products occurring in engineering and technology processes.

The technosphere can be defined as a system of relationships between man and nature , in which technology acts as a middleman and source of formation of a certain type of interaction viya - between technology - as an artificial environment, and nature - as natural environment.

Technique - 1) is a set of means of labor, tools, with the help who create something; 2) these are machines, mechanical tools, allpossible devices; 3) it is a body of knowledge, means, ways bobs used in some business.

The technique can be interpreted:
-               as a set of technical devices - from individual productsthe most sophisticated weapons to the most complex technical systems;
-               as a set of various types of technical activities,to create these devices

and systems;

-          as a set of technical knowledge, scientific and theoretical, andalso systemic.

The development of technical knowledge must be considered in a single together with the progress of natural science knowledge. However, it should be take into account the relative independence of the development of technical scientific knowledge and its conditionality by the progress of natural science and technology.

There are four stages in the development of technical knowledge.

Stage I. Pre-scientific (from the primitive communal system and ending the Renaissance), when technical knowledge existed as an empirical description of the means of labor activity and their methods application.

Stage II. The beginning of the use of scientific knowledge (from the second half guilt of the 15th century to the beginning of the 19th century), when for solving practical tasks begin to use scientific knowledge. At this time, continued the formation of natural science.

Stage III. Covers the period from the 19th century to the middle of the 20th century. it the so-called classical period in the history of natural science and technical sciences, when stable relationship.

Stage IV - from the middle of the XX century to the beginning of the XXI century. In this period there is an integration of technical and natural science knowledge, to which social and humanitarian knowledge is beginning to be connected niya. This is due to the emergence of automated

production, development of informatics, astronautics, new technologies giy - information, biotechnology, nanotechnology and others.

thinks, by analogy with the biosphere and noosphere, the field of technology is techno sphere.

The technosphere is a system of relationships:

-          between man and nature;
-          between technology and nature;
-          between man and technology;
-          technology and foundations of civilization and culture.

There is an opinion that the technosphere now occupies not only a priority place, but also enslaves a person, subordinating him to the law us of their evolution.

It must be borne in mind that human society has not yet learned moose to adequately use the achievements of technical progress , since psychological and cultural processes are slower scientific and technical .

1.4. Technology

Technology (techne - art, craftsmanship, skill, and logos word, knowledge) - a set of techniques and methods of obtaining, education handling or processing of raw materials, materials, semi-finished products or products li, as well as information carried out in various industries industry, construction, manufacturing, education, etc. Also called technology is a scientific discipline that develops ing techniques of any kind of human activity .

The development of technological knowledge must be studied in unity with

the progress of natural science knowledge. 1.5. Engineer

Vom began to be called a person engaged in the construction of anyThe word "engineer" has come into use worldwide since the 19th century. This word structures, devices, bridges, machines, devices, etc. in the as a leader. In the old days in England, such a person was called

captain; in France - the master; in Germany - a maester (master), and in Russia these - thought (meaning).

Reason was obliged to reflect the task assigned to it from all sides, relying not only on their own experience, but also on experience, captured by his predecessors, on his mind, ingenuity, dream, fantasy.

Thus, the Russian name "rassl" in its essence anticipated that understanding of the role of the leader in resolving technical scientific problems, which was established at a much later era - in the XIX century - and embodied in the concept of "engineer".

Excavations of ancient cities, fortresses provide a lot of evidence the existence of engineering (business of ideas) in a very distant times - XII - XIV centuries. For example, the first floating bridge in the world across the Dnieper near Kiev, mentioned in chronicles in 1115

year, was built under Vladimir Monomakh, and the wooden bridge The Volga cut was built under Dmitry Donskoy - in 1380.

Science, technology, art - various areas of human creativity century.

Technology stands to some extent between science and art, as and the engineer to some extent unites both the artist and the thought body.

1.6. Energy, energy, electric power

The material life of humanity is associated with two main principles - matter and energy. Therefore, all technical creativity humanity at all stages of the development of society was reduced, in essence state, to modifications and transformations of both matter and energy.

Energy (gr. Energeia - activity) - the ability of bodies (creatures) do the work. This action, a general quantitative measure of different forms of motion of matter. Energy connects all phenomena together nature.

Energy, energy science is the science of laws processes and phenomena directly or indirectly related to the receipt, transformation, transmission, distribution and use of personal types of energy.

The electric power industry is generally regarded as a complex technology education, closely interacting with the fuel economy vom and the main industries of the mining and processing industry thinking, transport, agriculture, etc.

Electrical energy is secondary energy and is not a substitute for it is primary, for example, thermal, hydraulic, wind, thermal nuclear, solar, tidal, nuclear, but at the same time, mimics their development.

The power industry is the leading energy sector. Applythe use of electricity, the use of electrical energy is one of the greatest discoveries and achievements of the 19th century. This was preceded by whether the efforts of many, many people. is the most convenient form of energy.   Now electrical energy is

The energy system of the electric power industry is called the aggregate power plants, electrical and heating networks, co-
united among themselves and connected by the commonality of the regime in unconditional disruptive process of production, transformation and distribution electric energy and heat with the general management of this mode mom.

16 sixteen

Energy is also a determining factor for the economy, and for the environment. The economic potential of the state depends on it.
va and the well-being of people. She also has the most powerful effect impact on the environment, ecosystems and biosphere as a whole.

Our planet is filled with energies that interact they live with it, with a person on its surface, with the Cosmos. Everything is energy gia! Spirit is energy, matter is energy, the smallest atom is energy gia.
Evolution brings high energies closer to the Planet, carrying and creative and destructive potentials. Which one is real lized - depends on humanity itself, on its energy sky, spiritual and moral potential ...
N.K. Roerich.

1.7. Scientific and technical progress
Progress - movement forward to a more perfect state; change for the better.
But does this always happen ?!
At the present time - at the beginning of the XXI century - a general, and especially scientific but technical progress has reached such a high level that, example, energy, electricity and industry provide
a very significant and far from positive effect on the biosphere ru and the environment due to the commensurability of capacities in the human systems and global processes in nature.
In his last article, one of the creators of the hydrogen bomb in Russia, academician Andrei Dmitrievich Sakharov wrote:

Scientific and technological progress will not bring happiness if it does not will be complemented by an extremely deep positive change the social, moral and cultural life of human state.

The inner spiritual life of people, their inner impulses activity is most difficult to predict, namely from this ultimately hangs and the death and salvation of mankind.

Despite the conservatism inherent in man, scientific technological progress cannot be stopped. Its main engine is there is work, the desire to make life better. History shows that people do not easily perceive new ideas, truths, and that, on the contrary,
they reject them the more stubbornly, the more these truths go beyond affairs of the sphere of feelings, habitual, ordinary.
We must understand and remember that labor is an inexhaustible source strength and consolation of a person in any situation.

Self-test questions

1. How important is knowledge of history (in a particular area)?

2. What are the goals and objectives of science? Its stages of development?

3. Under the influence of what factors does science develop?

4. What are the basic laws of the development of science?5. What role does technology play in social development?

6. What stages can be distinguished in the development of technology?

7. What is the importance of technology in human activities?

8. What is the role of electricity in the life of a humansociety?

Since the universe has existed there is no one who would not need knowledge.

What language and age we do not take, man has always strived for knowledge.

2. HISTORY OF ENERGY

2.1. General energy

Since ancient times, people have needed strength, engines, which which would help to uproot trees, would activate

devices for supplying water to the fields, plowing the land, rotating millstones grinding grain, etc.

In the countries of the Ancient East, in Egypt, India, China, for this whether already in the 3rd millennium BC. animals and slaves were used. Then on replacement of a living engine lam came water wheel - two discs per one shaft, between were placed planks - blades.

The flow of water in the river yeah pitchfork on the blades, turn

working wheel, and through   Fig. 1. Water Mill wheel shaft movement transmitted by millstones (Fig. 1).

In the 3rd millennium BC. people use used sails to move boats, but

only in the VII century. n. e. the Persians invented the wind a windmill with wings (Fig. 2). Beginning the history of wind turbines.

Water wheels have been used on le, Euphrates, Yangtze for lifting water, slaves rotated them. Then the ancient Greeks and the Romans used water wheels in as a motor for driving pumps and mills to squeeze out oil. Later water wheels have become widely used

Fig. 2. Windmill to study in a craft, then in industry.

Roman writer Marcus Vitruvius Polion in the 1st century. BC e. first described

sal water wheel. Water wheels and windmills all the way to The 17th century were the main types of engines.

In the late 17th - early 18th centuries in Italy, France, England, Russia, Spain and other states repeated attempts were made to create give a motor independent of the motor rippling river water and wind. The idea of using using steam to create an engine body arose thanks to reflection pits and experiences of ancient thinkers. Archimedes (c. 287 - 212 BC) (Fig. 3), one of the brilliant researchers bodies of the ancient period, the creator of the ancient her mechanics, great mathematician. From-

He left 5000 s of scientific and technical descriptions, tezha, sketches: sluice gates with sashes, textile machines, roller bearings, centrifugal pump, steam cannon, pistol with wheelsshutter, hydraulic press, mechanisms that convert returntranslational movement into rotator noe and vice versa, and much more. Giambattista della Porta (1538-

1616) investigated the formation of steam from water, which was important for further using steam in steam engines, researched the properties of a magnet.

The engineer de Cau in 1615 described steam devices for lifting water.

Figure: 5. Leonardo da Vinci

Otto von Guericke (1602-1686) post-

pitchfork and described experiments demonstrating the force of atmospheric pressure on the "Magdeburg hemispheres", from which air was removed, and This was achieved by using steam condensation. To sever these hemispheres, use used by eight horses.

Denis Papin (1647-1714) built the first technically implemented bathroom steam atmosphere tive machine representing a steam boiler in the form of a cylinder ra with a piston that rose

with the help of steam, and descended under the action of atmospheric pressure. Figure: 6. Scheme of the Severy pump: The cylinder was both a boiler and a worker 1 - cooling vessel; 2 - boiler;

mechanism at the same time.   3 - connecting pipe;

4 - crane; 5 - delivery pipe;

Thomas Savery (1650-1715) composed $_6$ - valves

gave a steam pump, in which steam-

the boiler was separated from the cylinder (fig. 6). Tsar Peter I bought a pump Severy for activating the fountains in the Summer Garden.

Thomas Newcomen (1663-1729) improved the steam pump, connected the piston with the balancer and the sump pump rod. Ohbed water was supplied to the cylinder from above to lower the piston (fig. 7).

Newcomen's machines were acquired retena by Peter I for pumping water from dock in Kronstadt.

Steam-atmosphere machines and Severi and Newcomen were cumbersome and had whether a small coefficient of efficiency consequences ($\approx 0.3\%$).

Ivan Ivanovich Polzunov (17291766), a talented Russian mechanic,

designed and built the first pa- Figure: 7. Machine structure machine with a universal heat Newcomen motor, (fig. 8). She had two cylinder with pistons and a separate steam boiler from which steam alternately entered the cylinders through an automatic distribution the divider is the first application of an automatic ticks in similar machines. Working effort continuously fed to the common pulley, whose shaft transmitted torque to the drive factory mechanisms - pump or air walking fur.

It was the first universal steam car, but still it had a small

Efficiency ($\approx 1\%$), consumed a large amount state of fuel; she worked for about a year in mines; after the death of the creator of the word Wasted and forgotten.

The first steam devices and machines

Figure: 8. Engine had low efficiency, since there was
diagram theoretical technical knowledge of h
I. I. Polzunova steam pressure and dr.

Mikhail Vasilyevich Lomonosov (1711-1765) - brilliant Russian scientist, thinker, experimenter, poet (Fig. 9).

Lomonosov did a lot in the field of various sciences and in each of them researched the most fundamental questions. He studied aggre-

state of matter, studied thermometry, introduced physical and chemical research methods. He experimentally proved and formulated in 1748 the law preserves the substance. It was 18 years before similar experiments of the Frenchman Lavoisier, who to which world science has attributed the discovery the law of conservation of matter.

Lomonosov for the first time gave the correct explanation of warmth as movement the smallest particles - corpuscles.

M.V. Lomonosov was not only you

talented and versatile scientist, but and passionate advocate of scientific knowledge. He understood the need for training for the people and gave it pain Your attention, remembering the testament of Peter I:

Figure: 9.M. V. Lomonosov

"... to produce and distribute sciences

wander. " Let us present Lomonosov's address in poetic form to to his students:

Oh you awaited
Fatherland from its bowels
And wants to see those,
What calls from foreign countries.
Oh, your days are blessed!
Dare now emboldened
Show with your hands
What can own Platons
And quick-witted Nevtons
Russian land to give birth.

About Lomonosov, the genius poet and philosopher A.S. Pushkin wrote: "Combining extraordinary willpower with extraordinary strength to understanding, Lomonosov embraced all branches of education. The thirst for science wouldla the strongest passion of his soul. Historian, rhetorician, mechanic, chemist, mineralogist, artist and poet - he experienced everything and in everything
Nick. "

Scientists, inventors, ingenious self-taught, mechanics continued we wanted to work on the device and improvement of steam tires and their application, having already some idea of heat.

James Watt (1736-1819), (fig. 10), English mechanic, created a double-acting steam engine, the working stroke of the piston in it was exhausted not by atmospheric pressure, but by pressure lazy steam.
Watt's car was driven by a golden device, (centrifugal regulator
rum steam). Contains a flywheel and connecting rod crank mechanism, made continuous jerky rotational movement. Condensation steam was produced in a separate device condenser. The overall efficiency of the machine was 8 %. In the second half of the 18th century. device the steam engine was worked out, she found wide application in industry
Fig. 10. James Watt   large countries. In honor of D. Watt, unit power was named "Watt".
In Russia, steam engines began to be built in St. Petersburg (at Galerny Island), Olonets and other factories.
American R. Fulton in 1803 installed a steam engine on the ship; such ships began to be called parokhodami.

Petersburg from 1800 to 1825 more than 100 steam factory machines and 11 steamboats. The first Russian steamer Elizave that "made flights" Petersburg Kronstadt "already in 1815.

Cherepanov Efim Alekseevich together with his son Miron Efimowich - mechanics of Nizhny Tagil factories - from 1820 to 1835 built or 20 different steam engines, and in

Figure: 11. Steam locomotive Cherepanovs

1833 built the first in Russia a steam locomotive, (Fig. 11), which moved along a cast-iron rail track.

The first railway in Russia "Petersburg - Tsarskoe Selo" was built in 1837.

D. Stephenson in England, starting in 1829, built a series of parovozov.

Created and invented various designs of steam machines, there was a need for the theory of both machines and heat transfer la.

The French scientist Sadi Carnot (1796-1832) in 1824 developed the foundations of the theory of steam engines are the Carnot cycles. He established that, the greater the temperature difference between the supplied and removed heat la at the coolant, the higher the efficiency of the heat engine. Co the time of S. Carnot thermal (steam, gas, etc.) machines of steel develop in the direction of increasing the parameters of the coolant temperature and pressure. These issues were addressed by R. Stirling, Erickson et al.

Water wheels and steam engines were improved, all were more introduced into industry, but they had a rather low cue efficiency and a relatively small

power. The creation of new machines with a large number of revolutions, with more power and greater efficiency. Such machines became various modifications of water, steam, and later gas

out turbines ("turbo" - top).

The theory of turbines was studied by D. Bernulli (1700-1782), who investigated dynamics of different energy flows.

In many countries, scientists, research vendors, mechanics offered various turbine design options. Was announced a competition for the best theory and best the design of the turbine.

B. Furneiron (1802-1867) constructed ruled a high-speed turbine with under-	
water on the blades radially from the center	Figure: 12. Turbine

turbine traction, (Fig. 12). Such a turbine is was widely used.	of Furneiron: 1-guide vane; 2-impeller blades; 3-shaft

Similar active turbines are developed personal kind were built by I. Safonov in Russia, Khovd in the USA, Girard in France, etc.

D. Francis (1815-1892) built a radial-axial jet tour bin with specially curved blades (Fig. 13), which received

	wide application nie. A. Pelton (1829 -1908) was created active bucket turbine for large the head of water . J. Poncelet (1788-
Fig.13. Francis Radial Axial Turbine (1) and Kaplan axial Kaplan turbine (2)	1867) created the turbines. It served as an impetus

to the creation of new types of machines.

Modern hydraulic turbines are based on selection and improvement of turbines built by many talented inventors and designers. The turbines were rotating under the action eat moving water. Then steam turbines appeared, in which superheated steam was used, supplied to the turbine blades under high blood pressure. The prototype of such turbines was eolipil GeRhone of Alexandria fig. 4. Steam turbines had a number of advantages over steam piston machines: fast rigidity, rotation uniformity, economy. Have appeared a variety of ideas and designs new turbines.

K. Laval (1845-1913) developed botal single-stage active

turbine with four steam nozzles, steam from which
fell on the turbine blades (fig. 14) , but using it it was economically
unprofitable although the principle is very valuable.
Fig. 14. Laval turbine

C. Parsons (1854-1931) art drifted a multistage axial jet turbine of large
power nost with special groups of blades - mobile and fixed.
This design was more successful and received further development.
the twist in the work of designers in many countries (France, England, Russia

these, America, etc.). The further development of steam turbines was
connected This is caused, among other things, by increasing the steam
temperature.
Steam engines and turbines required a device in which there would be a
firebox, a boiler, a cooling unit. They did their value, however, were very
cumbersome and inconvenient to operate.
Already at the end of the 17th century. the idea of creating an internal engine
its combustion is an internal combustion engine , in which a boiler and a
firebox are not needed, since gas the shaped working body receives energy
from the combustion of fuel inside working cylinder.
In internal combustion engines, the main part is a cylinder with piston, but it
is not steam that presses on the piston, but hot compressed gas, developed as
a result of fuel combustion inside the cylinder - fromhere and the name of the
internal combustion engine - internal combustion engine.
The first attempt to create an internal combustion engine was based on the
idea of H. Huygens
(1629-1695) - powder machine. However, it was not built, so as at that time
there was still no suitable fuel. In subsequent years, many models of various
internal combustion engines were developed, but all of them for one reason
or another were not implemented.
French mechanic E. Lenoir (1822-1900) invented the horizon double-acting
internal combustion engine. He RA-
worked on a mixture of lighting gas and air, had an efficiency of about 4%
and required good cooling. Lenoir's engine got satisfied
but high distribution, although it was far from perfect and demanding
I was in for serious improvements.
The first four-stroke engine the internal combustion engine was built by the

German Nikolai Otto in 1876, then he was improved promoted by a Russian engineer O. Kostovich, who developed lung combustion carburetor fractions of distillation products of oil ty. These same questions are fox German inventors DimeLehr and Benz (founders of the
"Mercedes").

Fig. 15. R. DieselGerman engineer Rudolph

Diesel (1858-1913) (Fig. 15), developed ICE on heavy fuel fuel oil, diesel oil. He worked on the principle of spontaneous combustion

27

27

niya. Internal combustion engines operating on the principle of self Ignition of fuel in the cylinder, called diesel, by the name of their inventor. The first diesel engine was manufactured in 1897 year, it contained all the basic elements of a modern motor, was the most economical of the internal combustion engine.

G.V. Trinkler - engineer of the Putilov plant, improved the process of fuel combustion, created in 1889 an engine with a mixed combustion, and from the beginning of the XX century. Nobel's plant ("Russian Diesel") became to start up diesel engines in Russia.

A great contribution to the development of energy, the creation of engines, running on organic fuel, scientists who discovered
and developing laws and the theory of various processes in the field chemistry and physics.

Dmitry Ivanovich Mendeleev (1834-1907) (Fig. 16) - issuescholarly Russian scientist, author of a fundamental periodical the horse of chemical elements, open which contributed to the development chemistry, atomic and nuclear physics Ki. DI. Mendeleev developed the combustion of fuel, which allows lala to determine the calorific value capacity of fuels of various composition, choose the optimal modes of niya and much more. Besides, DI. Mendeleev developed an industrial tailored methods of oil separation according to fractions - gasoline, kerosene, fuel oil, opened and formulated the position about the critical state of matter and much more. He was versatile

him a scientist, a patriot of his country Fig. 16. D. I. Mendeleev us, the propagandist of scientific discoveries
tii, professor of St. Petersburg University and other institutions

niy. DI Mendeleev's textbook "Fundamentals of Chemistry" (1868) was reprinted many times and is one of the best chemistry textbooks.

The work of scientists contributed to the development of progress, industrial laziness, energy.

28

28

In the twentieth century, a turbojet engine and a gas turbine bin. The development of such engines was initiated by an Englishman D. Barber back in 1791, when he received a patent for a heat engine a heater in which the combustion products of a mixture of air and gas were supplied on the turbine blades.The first working gas turbine engine was designed van and tested in 1897 by a Russian inventor engineer P.D. Kuzminsky (1840-1900), the fuel for this engine was kerosene; in the same year he built a gas-steam turbine with constant combustion pressure.

Work on the creation of turbojet engines, gas turbine bin were conducted in Germany (Stolz), in the USA (Moss), in France (ArMengo), in Russia (N. Gerasimov, V.I. Bazarov and others).

However, the construction of this kind of engines and their long-term work required heat-resistant materials and the development of a theory of gasof gas turbines. These questions, as well as the creation of highly effective tive compressor required for these engines,

in England, Germany (Heinkel's firm), the Soviet Union (A.A. SabLukov, B.S. Stechkin), France, Italy, Switzerland and other countries nah.

Gas turbine engines are widely used in aviation, steam and gas power plants, etc.

After various kinds of engines were invented wind, water, steam, turbojet, internal combustion

-        the question arose about the transmission of energy over a distance.

The programs were invented very different - belt (with the help belts), hydraulic (using fluid), pneumatic (with using air, gases). All of them could transmit energy, but on short distances and with significant losses. Development of mentality, construction of factories, factories, growth of large cities demanded more and more energy and its transmission over long distances.

The most important stage in the development of the energy base of industrial ness, agriculture, household amenities was the invention and the use of electric motors.
Electric motors are more convenient and reliable than other motors
-         steam, wind, water. They are always ready to go, they can controlled from a distance, allow you to adjust the speed, etc.

29
29

Thanks to electric motors, there appeared: high-performance machines, machine tools, automatic factories, electrified tool, electric transport (electric trains, trams, metro,
trolleybuses), household appliances (refrigerators, washing machines us, sewing machines) and much more.
The discovery of electricity and the use of electrical energy was one of the greatest events. This was preceded by efforts many, many people, from ancient times to the present day.
For long distance transmission and distribution between consumers - the most convenient is electricity sky energy.
It is believed that there is no useful electrical energy in nature, although there are such electrical atmospheric phenomena as molnii, northern lights, have electrical charges, some dwellers such as electric eel, electric ray.
steam and gases, has been used for a long time and continues to useEnergy of moving water, wind, energy of fuel, producing be a man. Installations, devices, moenergy consumption increases. This is due to the the need to improve methods of using energy sources and search for new renewable sources of nature.
The increase in human energy consumption in a number of cases is due to leads to harmful net effects of energy production on environment. This applies to organic fuels - coal, oil, fuel oil, gas, which, when burned, pollute the air, water, soil; this also applies to nuclear fuel that pollutes the atmosphere radioactive emissions and requiring for their radioactive
waste from the construction of special burial grounds for long-term storage niya. As a result of all this, humanity is increasingly focusing on increases the use of solar energy - solar energy, energy sea tides and biological energy,

which is implemented as a result of processing organic waste - biomass, total the mass of which is approximately 3.2 billion tons per year.
In the further presentation, we will consider the history of the appearance of electric energy development and development.

Self-test questions
1. How important is knowledge of history?
2. What is the material life of society connected with?
3. How important is energy in people's lives?
4. What is the energy system?
5. The first engines artificially created by human hands.
6. The first engines used by man are several thousandyears ago.
7. Names of ancient thinkers and inventors, their works.
8. The first inventors of steam engines.
9. Works of M.V. Lomonosov and their importance for world science.
10. The first inventors of steam and gas turbines.
11. The works of D.I. Mendeleev and their meaning.
12. The first inventors and developers of internal motorscombustion.

3. HISTORY OF THE DISCOVERY OF ELECTRICITY

Each generation finds the technique at the level to which it was brought in the previous period, supplements it with its own coverings, inventions, devices, and then transmits the following to the next generation.
The use of electricity and the use of electricity were the great discovery of the 19th century.
It should be noted that electrical energy is secondary energy and does not replace the primary (thermal, hydraulic, water dyana, etc.), but stimulates the development of primary energy, and for it transmission and distribution - the most convenient is the electric tric energy.
Electricity is highly concentrated energy:
$1 \text{ kWh} = 1000 \text{ J}/\text{s} \times 3600 \text{ s} = 3600000 \text{ J};$

1 kWh = 102 kg.m / s × 3600 s = 367000 kg.m - this is equivalent lifting 367 tons of cargo to a height of 1 meter.

The development of the power industry is international in nature. We are convinced of this throughout the history of its development. IN the creation of energy and its implementation have been strong participation of people of different nationalities, different countries, different classes.

For example, the first discoveries, scientific and practical developments ki, the laws in the electric power industry were the contribution of Italians, vat, Russians, French, Americans, Hungarians, Belgians, Yugoslavwws, Danes, etc. This can be seen by considering the history development of the electric power industry.

Wide and varied use of electricity in all areas of the national economy and everyday life is explained by a number of very its significant advantages over other forms of energy gii, namely: 1) the possibility of economical transmission on a significant solid distances; 2) ease of conversion to other forms energy (thermal, mechanical, light, chemical, etc.); 3) simplicity of distribution of any power (from many kilowatts to microwatt) between any number of consumers.

The ability to use for production is of great importance. electricity supply from local fuels (coal, peat, oil shale), energy of rivers, waterfalls, tides, solar energy and wind energy ra, geothermal, nuclear, etc.

However, both earlier and now there are numerous major problems of the electric power industry:

- creation of economical constructive sources of electricitystate - generators, electric motors, transformers, factories electricity (power plants), electric transmission lines
(Power lines), substations, switchgears;

- laying of conductors, cables, their protection;

- insulation of live wires, parts of devices;

- methods of calculating power grids, their protection against short circuits;niy;

- other issues that have been and are being solved by scientists,nera, practitioners, inventors.

3.1. History of discoveries in the electric power industry

The discovery and use of electricity was one of the greatest their achievements of mankind. This was preceded by the efforts of many and many people of different professions in different eras. Let's try list in historical order some of the most known discoveries, inventions, examples of the use of electricity wa and remember their creators.

In ancient Greece at the turn of the VII-VI centuries. BC. merchant, philosopher and

the scientist Thales of Miletus rubbed a piece of

32

32

nevshey resin - amber, which then received the ability attract various light objects: feather of a bird, dry leaflets, etc.

Many centuries later, an elementary charged particle (not existent unit electric charge) began to call the electron (in Greek - amber).

In the V century. BC. near the ancient city of Magnesia (territory with temporary Turkey) found amazing guiding oblong dark stones. They, suspended by long threads, always indicated one direction. These were lumps of magnetic ore, which later it was named after the city where it was found.

The first information about the use of electricity for metallization vessels belong to the III century. BC. (craft application). Used

The electrodes were made of copper and iron, and the electrolyte was wine. The electromotive force of such a chemical source of electricity reached ~ 0.8 V.

Then these discoveries were partially or completely lost (or forgotten), humanity invented and discovered them again.

The founder of the science of magnetic tism is the Englishman W. Hilbert (1540-1603), (fig. 17). In 1600 came out the work of W. Hilbert "On the magnet, magnetbodies and a large magnet - the Earth ", in which he describes the different poles near the magnet (north and south), lead the presence of identical and opposite fields owls, ways of magnetizing iron.

He was the first to indicate the presence of a magnet

Figure: 17. W. Hilbertfield of the Earth, devoting to this covering 18 years of life and putting about 600 experiments, he created the first electrical measuring device - electroscope and named electrical bodies capable of electrifying.

33

33

The first source of electricity already in our era was electric trostatic generator (triboelectric) invented in
1663 Mayor of Magdeburg Otto von
Guericke (fig. 18).
He made a ball of sulfur that rotated manually (by rubbing the surface hands). As a result, the ball accumulates an electric charge was generated. Power ball was less than 1 W. Seemingly a trifle, but with his help they were covered many important phenomena and properties electricity supply.

In 1675 I. Newton described the electric triization of bodies.

One might get the impression that Fig. 18. Otto von Guericke
The 17th century contributed little to the development of the science of electricity, but but then its foundation was laid and a powerful impetus was given to various figurative research of electrical phenomena of the following tables tiy.

F. Hawksby in 1705 created an electric generator using glass instead of a sulfur ball. In 1743, such a machine was introduced den sliding contact, which removed the charge, and the machine was able to rotation to continuously give off electrical energy.

S. Gray in 1729 noticed that some substances conduct electricity state, while others do not.

C. Dufay at the beginning of the 18th century. discovered electrical interaction charged bodies - the attraction of opposite and repulsion of the same registered bodies.

In the middle of the XVIII century. the "Leiden Bank" was established in Leiden
- a prototype of an electric capacitor. The opening of this condensation the torus belongs to the Dutch physics teacher Muschenbruck and to the German

priest von Kleist. Charged "Leiden Bank" with a sulfur ball von Guericke. "Leiden Bank" was a glass jar with a tatami of mercury on the inner surface. Insert it through the plug a nail was hit, and the outside of the jar was wrapped in metal foil. The nail and foil served as electrodes, and glass (dielectric) was deposited the charge from the sulfur ball of Guericke was pouring.

34

34

The experiment with a charged "Leyden jar" was demonstrated in the presence of a large crowd of people in a square in France. 180 of the king's guards stood in a circle, holding hands. One of guardsmen touched the foil of the Leyden jar, and the last in a chain touched a metal rod. All along the chain of guards deyets instantly flowed current and all people received electrical a blow that immediately caused a reaction of people - screams, jumping, waving hands, etc. Scientists have recorded the effects of electricity per person, the conductivity of the human body, as well as electricity sky blow.

Trying to charge the "Leyden jar" from heavenly electricity

VA (lightning), in 1753, Comrade M.V. Lomonosov G.V. Richman died.

Mikhail Vasilyevich Lomonosov, the founder of the Russian science, in 1753 he set a task for scientists: "... to find authentic the cause and make up its exact theory ".

M.V. Lomonosov did a lot of "celestial electricity" described electrical phenomena and the method of obtaining electricity artificially - the work "On electrical power ...".

With their friend G.V. Rikhman, they did a lot of observations and experiments with celestial electricity - lightning, northern radiance. Lomonosov expressed a very important thought about the possibility of transmission of electricity over long distances and practical the use of electricity for metal lization of the surface of metals (1747); onlyto 100 years later B.S. Jacobi opens and changes electroforming.

Georg Wilhelm Richmann (1711 - 1753) created a laboratory in St. Petersburg for research electrical phenomena, made a price a number of electrical measuring instruments.

In parallel with M.V. Lomonosov dilated experiments with "celestial electricity" in America B. Franklin (Fig. 19), scientist, poet,
Fig. 19. B. Franklin a diplomat who made a great contribution to the study the phenomenon of electrical phenomena and in 1752 invented a lightning rod (or rather I would call it a lightning rod).

35
35

The lightning rod from ball lightning was invented by a Russian engineer ner B. Ignatov in the XX century.

In 1759, the academician of the Russian Academy F. Epinus (1724 1802) discovered and explained the electric polarization, the existence of the development of magnetic field lines, the interaction of electrical and magnitric masses.
From the above, the vital an important conclusion (which we rarely think about): the first and very important discoveries in any field of knowledge are often made specialists from other branches of science or activity.
Let us confirm this statement with some more examples.

Italian Luigi Galvani (1737-1798) (fig. 20), manager
Department of Anatomy, in 1791 published the work "Treatise on the forces electricity during muscle movement nii ".
He discovered the existence of electric tric currents inside living
substances, dissecting with an iron scalpLem lying on a copper dish gushka (different metals!).
This discovery 121 years later gave push human research organism using bioelectric
currents. organs in the study of their electricSick people were found
signals. The work of any body   Fig. 20. L. Galvani
on (heart, brain) is accompanied by biological electrical signals that have their own form for each organ. If the body gets sick, the signals change their shape, and when comparing "healthy out "and" sick "signals, the causes of the disease are found.

Galvani's experiments prompted the invention of a new source electricity professor Alessandro University of Tessin Volta (1745-1827) (Fig. 21).

In 1800 A. Volta announced to the Royal Society of London woo about the invention of the voltaic pillar. He's his source of electricity
named after Galvani galvanic element. It was electricity source more more powerful than the Guericke generator.

This source consisted of a large number of small elements, each of which contained two plates of a pair different metals: copper - lead or silver - zinc, between which red was porous, fed with acid (or alkalichew), gasket.

Figure: 21. A. Volta   Typing   consistently
a large number of such elements

cops, Volta received an electrochemical source of electricity voltage up to 2 kV. This was already enough for research electricity, receiving an electric arc, an electric arc candle, welding metals, etc. A. Volta at this time was 56 years old. Like

Leon for this discovery presented him in 1801 with the Great Gold Medal. The batteries we currently use in watches, receivers and others are the same, but improved, voltaic columns - galvanilla elements.

In 1821, another source of electricity was invented - termy element. Professor T.I. Seebeck (1770-1831) discovered that if one junction of two dissimilar metals A and B (for example, copper constantan) heat (T g), and cool the second junction (T x), or simply not heating, then a thermoelectromotive force arises. nk

$$U_{AB} = \ln A \left(T_r - T_x' \right) qn \qquad_B$$

where k is the Boltzmann constant, q is the electron charge, n is the density electrons.

J. Peltier (1785-1845) discovered the opposite phenomenon in 1834

year. live a constant electrical voltage, then one of the junctions willIf one of

the junctions of two dissimilar conductors is applied be heated and the other

cooled.

$$Q_p = PI, \tau$$

where Q P - Peltier heat, P - metal pair coefficient, I -
current, - time.τ

With the invention of every new source of electricity, scientists discovered with interest that a mysterious electricity appeared under the influence of completely dissimilar forces, for example, heat, chimechanical reactions, mechanical friction, light, etc. Only penetrated innovation in the structure of matter, in atomic and molecular nature matter later made it possible to understand what unites these so different personal external phenomena.

3.2. The first laws of electrical engineering

Let's trace how the basic laws of electricity were established. This happened in a very peculiar way, sometimes by analogy with others
phenomena, sometimes speculatively, through the efforts of the thoughts of various scientists in different countries. Correct views on the nature of electricity whether your way gradually. Without fully understanding the reasons physical phenomena, researchers have established patterns, formulated laws.

The first important law of electricity was established by the French physicist Charles Coulomb (1736-1806) in 1785, long before inventions of galvanic cells. The wording of this law resembles the law of gravity - the force of interaction two point stationary charged bodies in a vacuum directly proportional to the product of their charges and inversely
is rational to the square of the distance between them. Attraction or frompushing off two charges meant their difference or their identity ness.

Physics school teacher Georg Ohm (1787-1854) discovered the law, which is very important: the current strength in the area is uniform electrical circuit is directly proportional to the applied voltage and is inversely proportional to electrical resistance growth of this area.

It should be noted that the XIX century. perceives an excellent tradition tion of the XVIII century. and leaves the memory of surprisingly versatile studies them.

Hans Christian Oersted (1777-1851) received the gold medal for
literary essay "The boundaries of poetry and prose" and at the same time pre-
staged work on the chemical study of the properties of alkalis. His his
doctoral dissertation on medicine and philosophy, he led research in the field
of pharmaceuticals. In 1813 Oersted publishes work on the effect of
electricity on a magnet, and in 1820 - work on direct my connection is a
magnetic effect and an electric current in a conductor the magnetic effect
arises around the conductor through which the current flows. Oersted's
publications encouraged other scientists and researchers to new discoveries
holes.

At a meeting of the French Academy, secretary D.F. Arago (1786-1853) (fig.
22) in 1820 reports on the discoveries of Oersted.

The mathematician André Marie Ampere (17751836) (fig. 23) present at the
report
Arago, the thought is born of the possibility of mutual mode of action of two
conductors with current. Outtoviv together with Arago from the salt guides
Figure: 22.D. Arago noids and passing a current through them. Ampere find-
dit that they behave like two magnets. Moreover, he is says the thought
(1822) that the magnet in in turn is a set currents.

Many years will pass and the discovery of these scientists and their names
will form the basis of the methods definitions will turn into names of a single
nits: electric current (ampere, A), kothe amount of electric charge (pendant,
Kl), voltage (volt, V), resistance (ohm, ohm), etc.

The creation of a magnetic field of electric current established experimentally
in 1820 J. Biot (1774-1862) and F. Savard   Figure: 23. A.M. Ampere
(1791-1841), and mathematically this phenomenon described by P. Laplace
(1749-1827).

Michael Faraday (1791-1867) (Fig. 24), originally had a the binder's
profession. He read a lot of articles intertwined by him

39

39

magazines and was very interested in the properties described there
electricity. Set up many experiments and subsequently did a number of
discoveries in the field of electromagnetnotism. His discoveries are at the

heart of the Denmark electric motors and power generators tori, transformers, electrolysis, optiand other phenomena. Faraday managed far ahead of their time. He proved that electricity and magnetism are inseparable connected. The phenomenon he discovered was chilo the name of electromagnetic induction; his description was published in 1831.
tia of electric and magnetic fields, ok-Faraday was the first scientist to introduce the concept$_{Figure:}$ 24. M. Faraday operating magnets and conductors with current. These fields are electromagnetic waves that propagate floating in space.

Faraday's discovery of electromagnetic induction refers to the most outstanding events of the XIX century. The work of millions of trans shapers, generators and electric motors worldwide based on the principle of electromagnetic induction.

Faraday's discoveries are based on research studies of previous scientists, and his research
ideas, in turn, prompted new discoveries of other scientists, engineers, investigators. Very often, scientists discover they embrace the new, looking back at the past.

M. Faraday's contemporaries English physicist D. Joule (1818-1889) and Russian scholar
E.Kh. Lenz (1804-1865) (fig. 25), one-
Figure: 25. E.Kh. Lenz   temporarily and independently of each other withdrawn thermal law
electric current - Joule-Lenz law:

$$Q \, d\text{-}l = 0.24 \, I^2 R \, \tau.$$

In 1832 E.H. Lenz established a law on the direction of induction current and formulated the principle of reversibility of generator and motor modes of electric machines. In 1845 he discovered and
described the phenomenon of the reaction of the armature of electrical machines. Lenz formulated

40
40

the most important position on the constancy of thermal conductivity and electrical conductivity of metals at the same temperature.

The discovery of electromagnetism led P.L. Schilling (1786-1837) (Fig. 26) to the isothe use of the electromagnetic telegraph in 1829 g.

The electromagnetic telegraph occupies fox: K. Gauss (1777-1855), W. Weber (18041891), B.S. Jacobi (1801-1874), S. Morse (1791-1872).

G.R. Kirchhoff (1824-1887) in 1845 onwrote a paper on the flow of electrical current through the flat plate and formulated shaft in this area are two fundamental Figure: 26. P.L. Shilling horse.

Decades later, James Clerk Maxwell (1831-
1879) (Fig. 27), developed Faraday's idea, clothed it in clear, precise material matic form. J.K. Maxwell created the mathematical foundation ment of the theory of electromagnetic interactions - four equations, four axioms that have not been subject to coopinion in the scientific world.

Below is a simplified view of these equations:

magnetic field strength1) Circular variation of the voltage vectorHξ generates electromotive force - electrical ξ

duction D   , and, therefore, a certain current density j : ξ 4π rotH

=   j (c is the speed of light). c

2)   Circular variation of the voltage vectorξ

        electric   field   strength   E   generates   Figure: 27.   J.C. Maxwell

$$\xi \qquad \xi \ 1\xi$$

 magnetic induction: B rotE = - B ...

c

ξ

3)   Electrical induction D depends on the dielectric value ε and the vector of electric field strength: ξ ξ

D = ε E ...

41   41

ξ

        4) Magnetic   induction   depends   on   magnetic permeability B ξ

μ and the magnetic field strength vector H :

        ξ ξ

$B = \mu H \dots$

The concept of the electromagnetic field took over the most important place in all sections of physics, electrical engineering, medicine.

Even in ancient times, doctors sometimes prescribed to patients treatment with "blows" of an electric ray in the water. In this way, it was possible to save the patient from paralysis. Around the excited nerve an electromagnetic field was generated. Electric "shocks" electric the nervous skate brought the excited nerve back to normal. Electric stingray and electric eel are living factories of electric current.

In Russia A.T. Bolotov (1738-1833) (Fig. 28) and I.P. Kulibin (1735-1818) (Fig. 28) created portable capacitive electric machines - "Leyden banks" for treatment patients and conducting psychological experiments Comrade Many years later, electromagnetic fields in humans, arising as a result of bioelectric signals.

Below we provide some information about wonderful Russian man Andrei Timofeevich Bolotov and other talented people. We do this in order to illuminate the action the hard work of some great workers

science and practice that have made an invaluable contribution into the movement of progress.

A.T. Bolotov wrote this about his life

Figure: 28. A.T. Bolotov   nom creed:

"Work without thinking about reward, do good without expecting gratitude. "
We list only the main works of A.T. Bolotova:
- he gave mankind the main principle of forest science and its use use, which underlies the entire global forest management

telii - to cut down the forest as much as possible to plant it; - published a work in 1766 on how to plant, grow and eat potatoes (at that time there was little about potatoes in Russia who knew), moreover, he himself started growing potatoes in order to feed the people who were starving at that time in Russia, and received a harvest

900 centners per hectare. French potato grower only after 17 years he set up his experiments with potatoes, got a harvest of 400 cent moat per hectare, and

for this only in France he was given four mint, and everyone forgot about Bolotov;

-       in 1770 A.T. Bolotov publishes an article "On the fertilization of the stranded ", where he writes:" All plants are composed of substances that belong reaping to the kingdom of minerals ... need for these things in the land and quantity to be ". Only 70 years later, in 1840, Liebig, who
ry is considered the father of agricultural chemistry, publishes his book, where he sets out foundations of the mineral theory of plant nutrition;

-       Bolotov - the founder of the science of apples - pomology ;

-       Bolotov was a doctor, herbalist and pharmacist - wasmedicines and successfully treated them. One of the first among healers, he used electrical devices created in conjunction with the talent the dashing Russian inventor Ivan Petrovich Kulibin for treatment of patients.

-       A.T. Bolotov was an outstanding journalist. His pen belongs there are numerous articles on agriculture, medicine. is he wrote a lot for children, he wrote the first children's scientific popular book "Children's Philosophy", which contains light denia in physics, mineralogy, botany, astronomy, cosmogony and dr.

The ability to think is the most important quality of a specialist, an engineer ra, scientist; find a pattern, be able to see, clearly form to formulate a problem and strive to find a solution is an integral features of a researcher's work.

The discoveries that people make are not them from above, they are the fruit of hard, persistent daily work.

We will present something very important for each person, especially in our time, statement of A.T. Bolotova:

"To be happy is a great skill! This is an internal state soul. And if labor and thought, the ability to rejoice at

every second of a fast-flowing life, a person will take possession of, and not spend life in search of material wealth or in search of happiness outside of oneself, then he will be able not only to live with dignity and joy, but also will make them happy. "

Everything in the world is interconnected! Studying inanimate matter, living organisms, establishing the laws of Nature, man at the same

43

time penetrates deeper and deeper into the innermost corners of the bya.

Ivan Petrovich Kulibin (Fig. 29) chief mechanic of the Petersburg Academy missions of sciences, a man of sharp, clear and technical of a sophisticated mind. In 1773 he created the famous project of wooden 1st single-arch bridge across the Neva (300 meters) from lattice trusses; the first put forward the idea of building bridges from iron forestry farms; created a navigable ship, self-propelled crew, optical telegraph and many other constructions tions.

Figure: 29. I.P. Kulibin There are a few things to say here.

words about Tsar Peter I.

Peter Alekseevich - Peter I (1671-1725) - the sovereign of Russia, surprised talented person. Sovereign educator, building the state leader, the "techie" ruler, who knew how to embrace his work zoom many different aspects of human activity, his life.

By order of Peter I in 1714, digital mathematic schools in a number of Russian cities. They were finished in his time I.I. Polzunov, K. D. Fedotov and other wonderful inventors.

Creating the Academy of Sciences, the sovereign took care that in it technical sciences developed; invited a capable

young talented scientists from abroad - the Bernoulli brothers, Euler and others; advocated for the spread of literacy among the people, for the development of domestic industry and crafts.

Knowledge of the history of technology, energy, science not only expands human intellectual outlook, but also has a great practical sky value, especially for a specialist. Introduction to diversity scientific and technical solutions of the past stimulate creativity activity, saves time and effort of an engineer, researcher, student. It makes it possible to use productively those inventions discoveries and discoveries that did not find practical application in their time changes due to the lack of need for them.

44

44

3.3. Initial period of electricity use

3.3.1. Electroplating, lighting and electrothermia

Electrotype. One of the first practical applications electricity was metallization - the deposition of a thin layer of metal on the surface of the product using an electric current.

This idea was expressed in the middle of the 18th century

M.V. Lomonosov, but applied practically through 100 years old, in 1847, B.S. Jacobi (1801-1874), (Fig. thirty). Since then, electroforming has become widespread to enter the industry. B.S. Jacobi, a talented engineer and scientist,

physicist, electrical engineer, invented and created electrical tric rotary motor em, created electroplating, several types electromagnetic telegraphs, applied electricity in mines, etc.

Figure: 30.B.S. Jacobi

Electric lighting - the first mass

energy use of electrical energy - has played

a key role in the development of the electric power industry and

the rotation of electrical engineering into an independent branch of technology. ElecThree-dimensional lighting was one of the first applications electricity after electroplating.

Vasily was at the origin of lighting with the help of electricity

Vladimirovich Petrov (1761-1834), professor of medical and surgery Chechnya Academy in St. Petersburg. He was the successor and continued Lem works by M.V. Lomonosov.

Investigating light phenomena caused by electric current, V.V. Petrov made his famous discovery - an electric arc, accompanied by the appearance of a bright glow and high temperature tours. This happened in 1802 and had a huge historical value. Observations and analysis of the properties of electrical arcs formed the basis for the creation of electric arc lamps, incandescent lamps niya, electric welding of metals and much more.

In 1803 V.V. Petrov was the first in the world to show the possibility of changes in electric current (electric arc) in metallurgy. Petrov research shaft electrical conductivity of various liquids and solids, expressed the idea of the

possibility of water decomposition by electric current, discovered oxidation and reduction reaction
metals, discovered the principle of accumulation electricity supply.
In 1875 Pavel Nikolaevich Yablochki (1847-1894) (Fig. 31), creates an electric candle consisting of two coal rods located vertically and parallel to each other friend, between which insulation from kaolin (clay). To burning (glow) was more efficient obligatory, on one candlestick
Figure: 31. P. N. Yablochkov placed four candles, which relied sequentially (in time).
In 1876, P.N. Yablochkova received recognition abroad; he becomes a millionaire. Streets of Paris, London theaters have become illuminated by "Russian light". Only after this Yablochkov candles began to be introduced in Russia.
Alexander Nikolaevich Lodygin (18471923) (Fig. 32), in 1872 proposed instead carbon electrodes in the Yablochkov candle is used use a filament (first coal-
nyu, and then from a refractory metal), whichparadise when an electric current flows brightly glowed. It was safe for people bright and cheap lighting by means of electric trinity.
A.N. Lodygin wrote that electric the light should be the only artificial light both in strength and evenness, as
Figure: 32. A. N. Lodygin and for safety and cheapness. List of inventions by A.N. Lodygin is very great. It includes electric induction and resistance furnaces, welding apparatus, batteries, electrical devices, extraction from ores

46

46

aluminum and other metals, an electric helicopter, a spacesuit and much, much more.
Dmitry Alexandrovich Lachinov (1842-1902) invented a lot various devices: voltage regulator, optical dynamo-
meter, method of centrifugal casting of reflectors. In 1880 D.A. Laranks wrote the book "Electromechanical Work", which contains sting research of the operation of electrical machines; it brought a mathematical proof is given

that over long distances it is possible to Any amount of electricity can be transmitted by increasing electrical voltage.

The issue of transferring electrical energy through wires to large distances were set for the first time in 1760 by M.V. Lomonosov; YES. Laranks and M. Despres conducted theoretical development of power transmission chi; F. Pirotsky and Fontaine performed for the first time or transfer using isolated wires waters and using conventional railways.

Thomas Edison (1847-1931) (fig. 33), talenta dashing American electrical engineer, an inventor who has his own ideas and the ideas of others quickly put into practice. They were improved Lodygin incandescent lamp was installed (pumped out air from a bulb, invented a base with Figure: 33. Thomas Edison screw thread, etc.); Edison's factories become whether to produce millions of incandescent lamps all over the world.

Alexander Ilyich Shpakovsky (18231881) creates in the 70s of the XIX century. arc lamp with electromagnetic and mechanical regulation, and in 1864 creates the first automatic steam pressure regulator direct action.

Vladimir Nikolaevich Chikolev (1845-
1898) (Fig. 34), creates a regulator for stabilization
lization of the combustion of an electric arc. He applied the arc light crushing system,
crushing the light of an arc lamp of 3000 candles into
60 light sources using the system   Figure: 34. V. N. Chikolev lenses, mirrors and tubes with reflective inner walls - light goods. With the help of such a device, Okhtinsky was illuminated

47

47

gunpowder factory. V.N. Chikolev improved the searchlights, changing the ring-shaped glasses and mirrors. He is the fundamental nickname of domestic lighting technology, the use of photography for opDetermination of the flight speed of projectiles and much more. Took active participation in the creation of the first power plants.

Nikolai Nikolaevich Benardos (1842-1905) applied electricity arc for welding metal sheets, cutting metals,
sti. Developed technologies for welding in an environment of shielding gases

and spotnoisy welding.

Nikolai Gavrilovich Slavyanov (1854-1897) (Fig. 35), created a constructures of electrical machines and apparatus, dynamos and regulators tori of an electric arc. I also used the electrode as a means for creating an electric arc, and as a metal carrier for creating seam when welding sheets or parts. He also applied electro heating of metal castings for uniform cooling throughout volume.

N.N.Benardos and N.G. Slavyanov used called the discovery of V.V. Petrov on melting ny and welding of metals in electricity arc.

One of the first Russian professors electrical engineering Mikhail Andreevich ShaTelen wrote:

"The first half of the 19th century. was special benno rich in the results of studying electrical tric current: was opened by electricity arc (V.V. Petrov), were discovered

Figure: 35. N. G. Slavyanov thermoelectric phenomena (T. Seebeck,

J. Peltier); found the law of thermal action of current (Joule's law

Lenz), the laws of the chemical action of the current (the laws

M. Faraday), the laws of G. Ohm and G. Kirchhoff were established, great clarity in understanding the phenomena of current; were discovered properties of current to magnetize iron and act on magnets; were

found the laws of interaction of currents with each other and the current with a magnetic tami; the laws of electromagnetic induction were discovered. "

With the discovery of the voltaic column, the current began to be used for various practical purposes: for lighting, for heating, for decomposition complex chemicals, for metal coatings and semi-

48

48

of metal impressions (electroforming of academician B.S.

Jacobi), for communication purposes (P.L.Schilling, B.S. Jacobi), for engines (E.H. Lenz, B.S. Jacobi) and others.

Over time, volts appeared other sources of electricity:

galvanic, thermoelements, dynamos, electric generators ry.

In addition to direct current, single-phase alternating current appeared, obtained from electromagnetic generators, and later - and threephase current

(M.O.Dolivo-Dobrovolsky).

3.3.2. The first accumulators of electrical energy

With the development of the electric power industry, its introduction into the industrial laziness, transport, everyday life there was a need to accumulate electricity
energy. V.V. Petrov at the beginning of the 19th century. creates the prerequisites for creating rechargeable batteries, conducts experiments.
G. Plante creates a lead-acid battery in 1859. K. Fore cona lead-acid battery was discharged in 1880 by A.N. Lodygin develops a theory of electricity storage for design operated electric helicopter.
In 1886 M. Despres creates a buffer storage battery.
In 1984, sodium sulfide batteries were created, much exceeding the technical and economic indicators of leadacidic.
Here are some numerical data for materials, method to accumulate electrical energy, per 1 kg of weight:

$$Pb \quad 16 \text{ Wh per 1 kg of weight}$$
Air - Zn 160 Wh —— "——
$$Li - Ni \quad 200 \text{ Wh} \text{ —— "——}$$
$$S - Na \quad 300 \text{ Wh} \text{ —— "——}$$
$$Li - Cl \quad 500 \text{ Wh} \text{ —— "——}$$
Petrol
$$engine \quad 2400 \text{ Wh} \text{ —— "——}.$$

The accumulation of electrical energy is necessary for working you are an autonomous vehicle - electric vehicles, electric helicopters, submarines; for energy storage during periods of low energy consumption and delivery of it during peak loads and in other cases yah.
To teach the strength of electric current to work miracles, generators and electric motors were needed. Many people thought about this.

49

49

some inventors, including Russians.
The future belongs to electricity! - was the conviction of the inventorlei. And who knows, don't open it then, at the end of the 19th century. rich and first pore cheap oil deposits, which gave the same rich and degasoline rivers,

could prevail in the competition with electric motor internal combustion engine ?!

Self-test questions

1. What are the advantages over other types of energy has electric tric energy (electromagnetic)?

2.From when and by whom were they discovered or used electrical phenomena?

3.   What are the first sources of electricity known, by whomLena?

4.   What are the names of the first researchers of the magnetic properties of bodies?

and electrical properties of charged bodies.

5.   What discoveries did the Leiden Bank help make? Who works Tal with her?

6.   What contribution did M.V. Lomonosov in research and application electricity? Which of the scientists worked at the same time in the field natural and artificial electricity?

7.   What discovery did L. Galvani make, and what did it push A.Volta?

8.   What sources of electricity appeared in the first halfXIX century?

9.   Who discovered the first basic laws in the field of electricity inXVIII-XIX centuries? What are these laws?

10.   What contribution did M. Faraday make to the practice and theory of electromagnetism?

11.   What contribution was made to the theory of electromagnetism by J.C. Mcswell?

12.   List the names of scientists, researchers, practitioners,at the origins of the practical use of electromagnet noisy energy.

13.   What are the names of the first Russian scientists and inventors, workersmelted in the field of electroplating, lighting, electrothermia.

14.   Who participated in the development of the first electric batteriesenergy?

3.3.3. Electric motors, generators, transformers

Discovery and research by D. Arago, G. Oersted, A. Ampere, G. Ohm, M. Faraday and other inventors and scientists served the impetus for the inventive imagination of engineers who became be called electricians. The most important stage in the development of electricity the invention and application of electrical machines was the first.

In technology, the main devices using the phenomenon electromagnetic induction are generators of electric

current, electric motors and transformers. Let's consider their main modern device and purpose, then to trace the history milestones in the development of these devices and indicate their authors.

Generator. Consists of stator and rotor . Massive nonthe movable stator is a hollow steel cylinder, on

the inner walls of which are laid a large number of turns of metal lined wire covered with insulation and carrying electricity into an external electrical circuit to the consumer.

The rotor is a cylinder with grooves for wires, which is emitted by a large movable electromagnet installed inside the stator.

or another motor, the rotor starts to rotate, and in the wiresUnder the action of a steam turbine, hydraulic turbine, steam engine

stator, due to electromagnetic induction, an electric cic current.

Electric motor. In electric motors, another phenomenon occurs n: an electric current flowing through the stator wires causes rotor to rotate. With the help of mechanical devices, the

the rotor can be transferred to the transmission belt, machine tool, escalator subway and other mechanisms.

Transformer. Consists of a magnetic core and two or more coils that have a different number of turns. If you let alternating electric current to a coil with a large number of turns

- the current of a higher voltage, then from the side of the coil with a lower the scrap of turns can be removed with more current, but less voltage.

Creation of electric generators, electric motors, transformers required the study of the properties of materials: nonmetallic, metallic and magnetic, the creation of their theory.

51

The first in this direction were the work of Professor Moscow
Skogo University of Alexander G. Stoletov (1839-1896)
(fig. 36). In the 80s. it was discovered hysteresis loop and domain structure ferromagnetic materials.

The Hopkinson brothers developed the Riyu electromagnetic circuits.

In 1895, Pierre Curie discovered a sous ferromagnets are critical temperature above which there is the disappearance of the domain structure and the loss of ferromagnetism is the Curie point.

Figure: 36. A. G. Stoletov The use of electricity for communication, lighting, motive power
the creation of electrical measuring instruments, Systems of units of measurements.

By the 80s. galvanometers, ammeters, voltmeters appeared, resistance stores, and the beginning of the creation of electrical instruments put M.V. Lomonosov, G.V. Richman, B. Franklin still in the XVIII century.

In 1881, the first International Congress met in Paris electricians. A resolution was adopted to develop a unified system we are units. The development group included: G. Helmholtz, G. KirchGough, W. Thomson, R. Clausius, A.G. Stoletov, etc.

Electric motors

The history of the creation of engines goes back to ancient times.

The man walked along difficult paths to the discovery and knowledge of the laws of physics. ziki, the creation of various mechanisms, machines.

For the first time, the engine was called a machine by the Roman architect Mark Polhe (1st century BC).

The most important stage in the development of the electric power industry was Breaking and application of electric motors. The principle of operation of the electric motor is based on a physical phenomenon: a coil of a conductor, to which an electric current flows when placed between magnets, moves across the lines of force of the magnetic field. Electhe trodden motor, as a rule, is more compact than other motors, always to work, can be controlled from a distance.

52

The history of the electric motor is a complex and long chain of discoveries, finds, inventions. Let's trace the stages of development of electric motors.

Stage I. The initial period of development of the electric motor (1821-1834 biennium). He is closely related to the creation of physical devices for demonstrations of continuous conversion of electrical energy into mechanical.

In 1821 M. Faraday, investigating the interaction of conductors with current and magnet, showed that electric current causes rotation of a conductor around a magnet, or rotation of a magnet around conductor. Faraday's experience showed the fundamental possibility building an electric motor.

Many researchers have proposed various designs of electrical trodmotors.

The first electric motors were similar in design to steam machines: engine by J. Henry (1832) and engine by W.

(1864) had rocker arms, a crank, a connecting rod, as well as spools (pecurrent switches in solenoids, replacing the cylinder).

P. Barlow proposed the "Barlow wheel". It consisted of constant magnet and gear wheels, sliding contact was carried out with using mercury, and the wheel was powered by a galvanic cell.

J. Henry proposed in 1832 a model of an engine with a reciprocating translational motion: moving electromagnet alternately attracted to permanent magnets and repelled from them, closing and disconnecting the batteries of galvanic cells. He did 75 swing niy per minute.

There were many more attempts to create rocking motors movement of the anchor. However, attempts to build a motor with a rotary motion of the armature.

Stage II. The second stage in the development of electric motors (1834-1860) characterized by designs with a rotary motion explicitly lusny anchor. However, the torque on the shaft of such motors the body was usually sharply pulsating.

In 1834 B.S. Jacobi created the world's first electric motor tel of direct current, in which the principle of direct rotation of the moving part of the engine. In 1838 this movement tel (0.5 kW) was tested on the Neva to propel a boat with passengers (Fig. 37), that is, received the first practical application nie.

53

Tests of various designs of electric motors have resulted in B.S. Jacobi and other researchers to the following conclusions: - the use of electric motors foris in direct proportion to the cheaper

the generation of electrical energy, i.e. from cosDenmark generator, more economical, than galvanic cells;

- electric motors must have possibilities are small dimensions and more power and more efficiency.

Stage III. The third stage in the development of electrical trodzig engines (1860-1887) is associated with development of structures with annular implicit pole armature and practically

Figure: 37. Bot Jacobi constant torque.

At this stage, the Italian electric motor should be noted A. Pacinotti (1860) (fig. 38). Its engine consisted of an anchor ring-shaped, rotating in a magnetic field of an electromagnets.

The current was supplied by rollers. Electromagnet winding connected in series with armature winding (i.e. electric the machine had a sequential excitement). Overall dimensions the engine were small, he had almost constant torque. In the engine le Pacinotti salient

the anchor has been replaced implicitly lusny.

Drum anchor in which   Figure: 38. Electric motor

rum worker is a wire   A. Pacinotti

the nickname constituting the coil was invented only in 1872 by V. Simen-catfish. After another 10 years, grooves for winding appeared in the iron of the armature

(1882). The drum armature of the DC machine became like this as we can see it at the present time.

The third stage in the development of electric motors is characterized by an open tie and industrial use of the principle of self-excitation,

54

in connection with which the principle was finally realized and formulated reversibility of an electric machine. Power supply of electric motors began to be produced from a cheaper source of electrical energy - direct current electromagnetic generator.

In 1886, the DC electric motor acquired the main features of modern design. In the future, he more and more improved.

By the type of current, electric motors began to be divided into machines for alternating and direct current; according to the principle of operation of the machine, Variable current are divided into synchronous and asynchronous.

Asynchronous motors are simple in design, low cost, reliability in work. They are the most common extended view of engines.

Power generators

Electric current generator prototype based on the principle zipe of electromagnetic induction, was designed by Faraday in 1831 It consisted of a copper disk rotating by hand between the poles of a permanent magnet. In this case, the disk was induced electromotive force (EMF); the poles were the axis of the disk and the a movable brush having sliding contact with the edge of the disc.

After that, various designs of electrical magnetic generators. Magneto-electric machines were manufactured produced by many inventors: W. Ricci, I. Pixie, J. Clark and others, but all of them were difficult to apply for practical use. vania.

By order of A.M. Ampere in 1832, I. Pixie (1808-1835) made the first electric generator with a commutator to obtain standing current. It was manually driven.

In 1842 D.S. Woolrich made a powerful DC generator current, connecting it with a belt drive with a steam engine. Such the generator was used to power the galvanic baths.

1842 is considered the year of birth of power supply acceptance.

In 1856-1866, the idea of self-excitation of electrogenic non-speaker (without galvanic cell). Many researchers, in-
The women, independently of each other, sooner or later came to this:
Hungarian A. Yedlik (1800-1895); German E.V. Siemens (1816-1892); English

Chane G. Wilde (1833-1919), S.A. Varley; American M.G. Farmer (1820-1893); Dane S. Hjerth (1802-1870) and others.

Industrial development of electrical trogenerators started after 1870

year, when the Frenchman Z. Gramm created heannular rotor rotor (Fig. 39), toroidal winding and number lecturer of an almost modern manuals. A. Pacinotti (1841-1912) at

10 years earlier built a similar

electric motor.   Figure: 39. Electric machine

 In 1880 the American T. Edison   Z. Gram

proposed to make a magnetic circuit

anchors of an electric generator made of insulated steel sheets Comrade This reduced losses and armature response.

In 1884, a compensation winding was proposed, and in 1885, additional poles to reduce armature reaction and improve commutation.

nom current solved many issues of the energy existing at that timeCreation of electric generators and electric motors for constant

tics, but the transmission of energy over long distances proved to be difficult body.

In 1876 P.N. Yablochkov created arc lamps, which are much worked more efficiently on alternating current. To supply several arc lamps from one source Yablochkov used induction tapped coils - a prototype of a transformer or The simplest transformer with an open core.

The introduction of alternating current was supposed to allow the transmission electricity using step-up voltage transformers

over long distances. But now the question arose about the creation of a AC non-speakers.

For the first time, the idea of a rotating electromagnetic field was expressed by D. Arago in 1821. In 1885 G. Ferraris. (1847-1897) suggested use a two-phase current (a system of two alternating currents, phase shifted by 90 °), which makes it possible to obtain "rotation an oscillating magnetic field ", and built an AC motor.

N. Tesla (1856 - 1943) (Fig. 40), managed to build a system of two-phase generator, transformer and motor.

It was used in Niagara hydroelectric power station in the USA, the system required four wires for the transmission of electricity gii.

In 1888, the Russian inventor M.O.

Dolivo-Dobrovolsky (1862-1919) (Fig.

41), created a three-phase system of currents, which toraya then received recognition and distribution ranked all over the world as the most

Figure: 40. N. Tesla convenient and economical.

The rotating magnetic field was obtained by phase shifting between currents of the same amplitude by 120 °. M.O. Topping up Dobrovolsky developed a rotor with a winding coy in the form of a squirrel cage and created a short closed-loop induction motor.

Three-phase system consisting of three-phase generator, three-phase motor gator (fig. 42), and a three-phase transformer mator, demanded for transmission and distribution division of electricity with only three wires, being at the same time symmetric, the equation innovative and economical. Costs of methallus were 25% less than two wire line of a single-phase system.

Figure: 41. M.O. Topping up

Three Phase Synchronous Generator   Dobrovolsky

was built by Dolivo-Dobrovolsky in

1890 For the first time, the transmission of three-phase current over a distance of 170 km was

la demonstrated at the International Electrotechnicalheadquarters in Frankfurt am Main in 1891 during the International

Congress of Electrical Engineers.

Based on electric generators and electric motors, individual drive of machine tools, mechanisms and devices.

57

57

The first protective grounding of electrical machines was proposed Russian engineer R.E. Klasson and the Frenchman M. Despres.

Electric current generators showed the following to the prime mover requirements: high speed, high kaya uniformity of rotation and continuously increasing power. Steam machine did not meet these requirements anymore, She had 400-600 rpm. Steam machine youoppressed by the steam turbine, which had higher speed and higher efficiency.

Now the capacity of steam turbines reaches

1200 MW. Turbine with electric

Figure: 42. Three-phase   a generator is called a turbo generator.
engine

Transformers

In 1848, the French mechanic G. Rumkorf invented the induction new coil. She was the prototype of the transformer.

P.N. Yablochkov, a Russian inventor, developed a system of lheniya "electric energy, for the first time using induction coil as an open core transformer for supplying several arc lamps. In essence, in 1889 he created the first power transformer.

In 1882, the Russian electrical engineer I.F.

Usagin, and in 1884 a French engineer

Bolard created a voltage transformer (to increase or decrease the tension). Development of power transformatters made it possible to transfer long distance electricity, so as with an increase in the value of the voltage losses are reduced electrical energy, and appears

the ability to reduce the cross section of the wire devs (Fig. 43).

Figure: 43. Transformer

In 1885 Hungarian engineers

M. Deri and O. Blati, together with K. Ziperovsky, developed a transform closed magnetic circuit mators. A distribution system appeared

58

58

power supply based on parallel connection transformers to the high voltage supply network.

Currently in power plants and substations use step-down and step-up, two- and three-winding, three-phase and single-phase power transformers (fig. 43).

Current transformers are used in AC installations all voltages for series measuring coils devices and protection relays.

The primary winding of the current transformer is included in the circuit for therefore, and to the secondary winding also sequentially connect the coils of devices and relays. Between primary and secondary there is no electrical connection with the current transformer windings, therefore they reliably isolate devices and relays from the installation voltage.

Voltage transformers are used in alternating installations current for supplying parallel coils of measuring instruments ditch and protection relay. Primary winding of voltage transformer connected in parallel to the network, and connected to the secondary winding coils of devices and relays are installed in parallel.

The transformer is one of the key components of the temporary energy system. It converts voltages into low or high with low energy loss. Is an important

element of many electrical appliances, mechanisms and devices: charge devices, radios, televisions, substations, electrical stations, etc.

Transformer sizes can range from a pea to hulks weighing 500 tons.

Reducing the size of transformers is achieved due to more efficient heat dissipation with fans, external radiators, special pumps. Evaporation systems are used cooling, but they are still too expensive.

Insulation and cooling system improvement process transformers continues: the design of transformers, cooling methods, we are looking for the possibility of using use of superconductivity windings.

Currently, the functions of transformers can be taken on bja semiconductor devices. However, transformers will still be perform their service for a fairly long time, effectively and discreetly supporting the functioning of electric power systems on which so many depend in our modern life.

59

59

3.4. Research and application electrical materials

Development of the electric power industry, manufacture of power sources energy, all kinds of devices for its use and

transmission over long distances, expanding the area of their practical its application required the development and creation of various electrical technical materials.

All researchers, inventors, electrical engineers understood in order to implement your ideas, you need to materialize them, and this requires materials with specific properties.

V.V. Petrov created one of the best laboratory in the world, in which there were more than 630 instruments in the swarm. Most of these devices he made the moat himself. In this laboratory, wonderful scientists E.H. Lenz, B.S. Jacobi et al.

In his laboratory V.V. Petrov tested electrical conductivity various solid and liquid materials, compiled reference tables faces of their properties.

Materials began to be divided into conductive and insulating. Current must flow through materials with high electrical conductivity, but at the same time, these conduction channels must be isolated apart.

Petrov was one of the first to insulate the conductors of the current lump soaked in resin or oil. Then he did isolation with using molten sealing wax.

In 1812, the Russian inventor P.L. Schilling suggested guttapermeable insulation for wires laid under water - underwater electric cable. In 1832, for an electromagnetic television Count P.L. Schilling used rubber and film impregnated with com. In 1837 he established an underwater telegraph line with rubber isolation between Petersburg and Kronstadt.

In 1839 B.S. Jacobi invented the writing electromagnetic television a graph whose wires he insulated with rubber and placed them in lead howling tube.

The first underground telegraph cables were proposed by P.L. Shielling and B.S. Jacobi. Insulated cable was passed through glass or steel tubes, which in turn were laid in wood Withered gutters placed in trenches.

By introducing sulfur into rubber, rubber was obtained. If the content the sulfur in the rubber was increased, then ebonite was obtained - a solid

60

60

for the manufacture of electrical appliances. In the early 40s of the XIX century. start They widely use rubber and gutta-percha.

In 1847, W. Siemens (Germany) used rubber insulator tion of wires and cables.

Since 1879, the insulated wire was covered with lead shell. In 1879 F. Borel (Switzerland) developed the technology manufacture of cables with a lead sheath. Since 1890 they begin introduce oil-impregnated paper insulation.

The development of electrical machines, generators has caused the need for development of heat-resistant insulation. Heat-resistant
impregnating compositions and coatings, composite compositions for isollation of plates of the collector of machines. As insulation steel used call natural mica (muscovite and phlogopite), then mycanites, micalents, micafolia, micalexa.

At the end of the 19th and the beginning of the 20th centuries. new materials are being created - syntethetic high molecular weight compounds with good insulating properties - polyethylene, polystyrene, vinyl plastic, polyvinyl
chloride, polymethyl methacrylate, etc. Later, high-temperature thermal polymers such as fluoroplastics and organoelement compounds, crosslinked and irradiated polyethylenes, etc.

At the end of 1906, checkpoints, support, suspension insulators based on glazed ceramics, later these insulators began to be made of electrical glass and special special plastics.

The use of such insulators was necessary to transmit electric current through the air over long distances, when required The voltage increased to hundreds and thousands of kilovolts.

Later, the problem arises of the transmission of large currents at relatively low operating voltages. To resolve these issues ows it was necessary to reduce the resistance of power lines.

You can significantly reduce the resistance of conductors by deep cooling them. Such conductors are called guconductors.

In 1911 the Dutch physicist Kamerling-Onnes discovered the phenomenon superconductivity, when the electrical resistance of the conductor kov becomes equal to zero (usually at 1 ... 8 K). Several dozen For years, superconductors could not be used in practice, since it was necessary to create very low temperatures to convert them to superconducting state.

61

61

After the 50s. XX century alloys were discovered (for example, niobium with tin), which passed into the superconducting state at higher temperatures.

High-temperature superconductors. In 1986 A. Müller (USA) and G. Bednorz (Switzerland) discover high-temperature superconductivity ceramic materials based on trivalent copper with a temperature the transition to the superconducting state 35 K and 78 K.

Improvement of electrical generators, motors, transformers required the study of the properties of metals, magnetic materials and creating alloys with high ferromagnetic properties estates. The solution of these problems was facilitated by the work of P. Curie, A.G.

Stoletov, the Hopkinson brothers, T. Edison and others.

Self-test questions

1. Research and work of which authors served to create the firstout electric cars?

2. Trace the main stages of creating electric motors and nasvite the authors of the developments.

3. For what purposes were the first electric motors used and by whom?

4. Name the first creators of electrical generators.

5. What designs of electric generators were proposed by the inventorlyami?

6. The creation of which generators allowed the transmission of electricityenergy over long distances?

7. Who developed multiphase systems of alternating currents, asWhich of them are widely used?

8. Describe the way of creating transformers of different types and namethose of their authors.

9. Who participated in the development and creation of electrical materialsterials?

10. What materials were used as conductors, and whatky as insulators in the manufacture of electrical products and transmission lines?

62

62

3.5. Electric stations

Power plants - factories for the production of electrical energy gii to be distributed among various consumers, did not appear immediately.

In 1873, under the leadership of the Belgian-French invention, body Z.T. Gram (1826-1901) the first power plant was built several kilowatts to power the lighting system of the plant, so called block station.

In the 70-80s. XIX century. every more or less respectable consumer tel (factory, street, house) had its own source of electricity (its power plant). As prime movers that lead to

the movement of an electric generator, piston steam was used at first machines, sometimes internal combustion engines or locomotives whether. A belt variable led from the prime mover to the electric generator.

transmission.

The first power plant (block station) in Russia was built at the Sormovskiy machine-building plant for supplying lighting installations in 1876

The first block station in St. Petersburg was built in 1879 at participation of P.N. Yablochkov to illuminate the Liteiny Bridge.

In 1879, the first power plant in the United States was built at San Francisco 30 kW under the leadership of C.F. Bran.

The first central stations appeared already in the 80s of the XIX century. They were more expedient and more economical than block stations, since they supplied electricity to many enterprises at once.

At that time, mass consumers of electricity were used light sources - arc lamps and incandescent lamps.

The first power plants of St. Petersburg were initially located on barges anchored at berths on the Moika and Fontanka rivers in chalet of the 80s. The capacity of each station was approximately 200 kW.

The world's first central station was commissioned in 1882 in New York, it had a power of 500 kW.

The first 400 kW central power plant in Moscow (Georgievskaya) was built in 1888. Steam heat was used.

rugged boilers of the Shukhov system, the main fuel was stone ny coal.

63

63

The first steam turbine at a power plant in Russia was installed renovated in St. Petersburg in 1891. All power plants initially worked on direct current, and this limited the service radius of the hitters several hundred meters, since the losses amounted to almost ti 20%.

An increase in the range of power plants could be carried out only when power plants are switched to alternating current, when which could use step-up transformers.

In 1884, an AC power plant was built in London.

Large single-phase AC power plant in Russia was built in 1887 in Odessa to illuminate the theater.

In Tsarskoe Selo, the length of the electrical network in 1887 was put 64 km. Tsarskoe Selo was the first city in Europe that The rye was lit exclusively by electricity.

The largest single-phase power plant in Russia at 800 kW was built on Vasilievsky Island in St. Petersburg in 1894.

under the guidance of engineer N.V. Smirnov.

The use of alternating current made it possible to simplify and reduce the cost electrical network. The cross-section of wires from 400-600 mm 2 decreased up to 58 mm 2 , and losses decreased from 20% at direct current to 3% on alternating current.

The beginning of the current stage in the development of the electric power industry is is set to the 90s of the XIX century, when the complex energy the problem of power transmission and electric drive - application three-phase current.

The founder of the system of three-phase alternating currents was the Russian scientist M.O. Dolivo-Dobrovolsky.

The creation of a three-phase system was the most important stage in the development development of electrical engineering and opened the way for her to industry.

The first enterprise in Russia with a three-phase power supply it was the Novorossiysk elevator (1893), the builder of the power plant tion was a Russian engineer A.N. Shensnovich.

The first three-phase plant in America was built in Cali-

Fornia at a hydroelectric power station in 1893

Electrification of the Okhta Gunpowder Factory in St. Petersburg a three-phase current system was implemented by electrical engineers technicians V.N. Chikolev and R.E. Klasson.

R.E. Klasson (1868-1926) led the construction of a number of power plants in St. Petersburg and Moscow, invented a hydraulic method sob of peat extraction, etc.

The creation of a three-phase system was the most important stage in the development vity of electrical engineering. Electrical energy from places of its cheap

receipt (river, coal deposits, peat) could now be transferred to remote industrial and urban areas. Electrification process katsii began to capture all new areas of production human capacity, contributed to the development of productive forces.

The electrification of large cities in Russia began in 1897.

At the end of 1906, suspension insulators were invented, which allowed to increase the value of the transmitted voltage.

The first energy system in Russia was created on the basis of two power plants in Baku. This system was created by electrical engineers techniques R.E. Klasson and L.B. Krasin in 1902

The electrification of Russia will be discussed in more detail farther. The next paragraph is devoted to various types of electrical stations that exist or are being developed at present me.

3.5.1. Types of power plants

Power plants that began to be intensively built in
70-80s XIX century. and continue to be built at the present time - i.e. in the beginning of the XXI century, can be classified by primary energy into the following types:
- thermal, including nuclear,
- hydraulic,
- wind,
- helio stations,
- geothermal,
- tidal, etc.

Thermal power plants, depending on the type of primary engine (and electrical energy is still only secondary energy), are divided into:
- in power plants with steam turbines, condensing IES or GRES (state district power plant), produced only electrical energy melts;

\-   with heating turbines - CHP, in which one partheat energy is given to consumers in the form of hot water or steam, and the other part is used to generate electricity;

\-   with piston machines - locomotives or diesels;- with gas turbines; - atomic, etc.

Let us briefly consider what the listed types are power plants.

Thermal power plants

They consist of the following main large nodes (Fig. 44).

Figure: 44. Scheme of a thermal power plant

1. Boiler plant - serves to generate steam from water for through the use of fuel heat (coal, peat, oil, fuel oil, etc.). The boiler plant includes a furnace, where the topliva, and a steam boiler. In addition, it includes a superheater, economizer - for heating feed water, air heating tel - for heating the air for the furnace. The boiler plant has also auxiliary equipment:

•   draft device - natural, in the form of a high pipe,or blowing fans - artificial;

•   ash collectors (filters);

•   water treatment - water purification.

2.   Steam turbine, (fig. 45). It consists of a shaft on which disks firmly mounted. On the rims of these discs are specially fixed curved working lobes molasses. The shaft rotates on bearings. Steam from boiler nozzles falls on working shovels turbine and forcing the turbine shaft rotates chat. Coming out of the turbine wanderings, steam condenses to create more re-

        Figure: 45. Turbine section   drop in pressure, accelerated

steam flow and increase cycle efficiency. Pressure decreases from 100 to 0.4 atm.

3.   Electric generator. Its rotor is on the same shaft with the turbine Noah. An electric generator generates electrical energy that paradise enters the converter station.

Gas turbine power plants

They work on approximately the same principle as the steam bin, but not water is used as a working substance
steam, and gas is a fuel combustion product or natural gas. Applythe modernization of gas turbines is currently one of the promising zheny costs for obtaining electric and thermal energy. For they are characterized by block-modular design, high technical characteristics, small dimensions (since no steam boiler is required and steam lines). To increase efficiency, they are complete are used by a waste-heat boiler, the efficiency with such a boiler is an example but equal to 77%.

67

67

Hydroelectric power plants (HPP)
Hydroelectric power plants use energy of moving water (fig. 46).

Figure: 46. Hydroelectric power plant

The prime movers of power generators are hydroturbines in which the potential and kinetic energy of water is is formed into mechanical energy to rotate the generator rotor.
The hydraulic turbine consists of an impeller and a guide swarms. The impeller is rigidly fixed to the turbine shaft, has on its rim, along the entire perimeter, a row of especially curved blades.
The guiding device gives the water movement the required direction

management and regulates the amount of water entering the turbine from using swivel blades. Guide reversal mechanism
blades is connected to the turbine regulator, which maintains a constant
the number of revolutions of the turbine and thus the frequency of the current. From the worker wheels, the water is discharged through the suction pipe into the tailwater of the guide rostations.

The power developed by the turbine on the shaft, P_t, depends on the flow rate water Q (m 3 / s), from the pressure of water H (m) (i.e., from the difference in heights of water I wait for the upper and lower reaches of the river) and from k - the coefficient of useful the action of the hydraulic turbine, $P_т = Q \cdot H \cdot k$.

The hydropower industry in Russia is characterized by a high degree of concentralization of capacities. The largest HPPs are shown in Table 1.

68

68

Table 1 The largest hydroelectric power plants in Russia

Power station	River	Installed power, MW	Average long-term design output electricity, bln kWh
Sayano-Shushenskaya	Yenisei	6400	23.30
Krasnoyarsk	Yenisei	6000	20.40
Bratsk	Angara	4500	22,60
Ust-Ilimsk	Angara	3840	21.62
Volgograd	Volga	2541	11.10
Volzhskaya	Volga	2300	10.90
Cheboksary	Volga	1370	3.31
Saratov	Volga	1360	5.40
Zeyskaya	Zeya	1330	4.91
Nizhnekamsk	Kama	1205	2.54
Votkinskaya	Kama	1020	2.32
Chirkeyskaya	Sulak	1000	2.43

| Zagorskaya PSP | Kunya | 1000 | 1.20 |
| Bureyskaya | Zeya | 2000 | - |

Figure: 47. Nurek HPP

...

In fig. 47 shows a diagram of the high-mountain Nurek HPP, height its dam is about 300 meters.

69

Geothermal power plants (GTPP) They use the internal heat of the Earth, geothermal energy

geysers, thermal springs for district heating and forproduction of electricity.

In Russia, geothermal springs exist in Kamchatka, Kuril Islands, Siberia.

For the first time a geothermal station on deep steam with a pressure of 5 atm and a temperature of + 200 ° C was built in Italy in the city of Larderello in 1904

Geothermal stations are used in Italy, Iceland, Russia these, Japan, New Zealand.

Pauzhetskaya was built in Russia in Kamchatka in 1967

GTPP for 11 MW, at the beginning of the XXI century Mutnovskaya for 200 MW, is building Xia Paratunkovskaya GTPP.

Solar power plants (GLES)

They use the thermal energy of the sun's rays using receivers of two types:

-              flat, catching the sun's rays, directed towardsperpendicular to the plane (receivers track the direction of the sun nechny rays, automatically unfolding its plane);

-              concentrating, in which the sun's rays with the help of mirspherical surfaces are concentrated in focus, where

thermal elements of the installation are located (for example, steam tel).

Solar power plants based on semiconductor new photocells (silicon, selenium, etc.). In such a setting New solar energy is directly converted into electricity energy.

At the end of the XX century. in the USA and Russia, a two-layer semia

conductive gallium arsenide photocell that converts
into electricity the visible part of the solar spectrum, and the infrared the part
of the spectrum passing through this transparent layer absorbs it is converted
and converted into electricity in the second layer - antimonide gallium or
aluminum arsenide. As a result, the efficiency of such a photocell is is
approximately 30-37%, which is comparable to the efficiency of modern
thermal and nuclear power plants (for conventional photovoltaic cells in
currently, the efficiency is somewhere around 10-12%).

70

70

In Italy, a solar power plant with a steam turbine has the power power 200
kW. A semiconductor solar was built in Armenia an unusual power plant
with a capacity of 1200 kW.

Tidal hydroelectric power plants (TPP)

Such stations generate electrical energy by using using the potential energy of
the ebb and flow of the sea.

The magnitude of the tide (as a result of the attraction of the moon) in
different places The land is not the same: off the coast of America it is 21 m,
off the coast of France and England - about 15 m, off the coast of Russia - 8
... 11 m on White and Okhotsk seas. Found to use energy tides are advisable
already at 3-4 m high tide.

Tidal stations are built in narrow bays. Perego-
The entrance is filled with a dam and hydrogenerators are installed in it. In
the time of the ebb and flow, the water flows through the pipes to the turbines
androtates them, and, consequently, an electric generator sitting on one shaft
with turbine.

For TPP, reversible turbines are used when rotation is not discontinuously in
any direction of water movement.

The tides to rotate the mill wheels were used 1000 years ago in Spain,
France, England.

TPPs operate in China, France, Russia (Kislogubskaya
TPP on the Barents Sea has a capacity of 1200 kW) and other countries nah.

Wind power plants (WPP)

These stations use wind energy. Wind energy is used has been used by

humanity for several millennia, but for the development electricity, mainly in the twentieth century.
Most often, vane-type wind turbines are made.
The diameter of the wings ranges from 8 to 30 m or more, and the power of such Tanks from 1 to 1000 kW and more (Fig. 48).
The power of the wind engine P determines the wind speed C, m / s and k - unit efficiency -
P = k . C 3 (kW).

Nikolai Yegorovich Zhukovsky (1847-1921) proposed a law on wind turbine wing structure.

71

71

There are tens of thousands of wind turbines in Russia, and their designs are constantly being improved.

During the period of full wind operation, electric tric energy is accumulated with the next return in a period of calm.

Russia at the beginning of the 20th century was one from the leading countries in practical use
low wind energy. In the 30s was
built in the area of Balaklava wind power
power plant with a capacity of 100 kW, with a wheel meter 30 meters. Its creation
associated with the names of scientists and engineers moat - G.Kh. Sabinin, N.V. Krasovsky,
V.R. Sektov and others. Then, in connection with
Figure: 48. Wind turbine construction of powerful power plants,
vane interest in wind farms has dropped. They are now are gaining strength again.
Pumped storage power plants (PSPP)
For the operation of such stations, it is necessary to have two reservoirs - top 2 and bottom 3 , located from

a friend in height by 40-600 m. Between the reservoirs pipes are laid through which it pumps water with a pump owls 1 (Fig. 49). In the hours of maenergy consumption logo, usually at night, free

Figure: 49. Hydraulic storage power station

electric energy station is spent on work that electric motors on-pumps that raise water from the lower reservoir to the upper water pre-storage.

During peak hours (high power consumption), hydroaccumthe power plant operates in a discharge hydro-turbine mode when water from the upper reservoir passes through pipes

72

72

to the bottom. Changing the mode from pumping to turbine, or vice versa mouth, occurs in 1.5-3 minutes. The efficiency of such stations is 70-75%.

Nuclear power plants (NPP)

Nuclear power plants use nuclear power to generate electricity and heat. fuel. Instead of a boiler unit, nuclear power plants use

Figure: 50. Scheme of a fast neutron nuclear power plant

there is a nuclear reactor and special steam generators (Fig. 50).

A nuclear power plant uses a substance capable of spontaneous fission of atomic nuclei with the release of energy in the form of heat.

The most important nuclear fuel is heavy elements: uranium-235, uranium-233, plutonium-239.

The fission of uranium-235 nuclei occurs under the influence of it thrones in a chain reaction, while a large number of thermal energy (~ 83%) and so-called nuclear radiation (~ 17%).

At a nuclear power plant (Fig. 50), the main one is the nuclear reactor

1 - a huge metal cylinder. Inside the reactor are rods with nuclear fuel - fuel rods (lower part of Fig. 50), grafit rods and pipes with water. The uranium rods are nuclei, and the graphite rods absorb "extra" neutrons and keep the fission reaction at the required level.

The energy obtained from nuclear fission heats water (or other
coolant, for example, metallic sodium). 10 drive the hot coolant from the reactor into the steam generators Pumps 3, 6, 8,   2, 4, 5.

73

Steam generator (superheater) device - pipe in pipe larger diameter. In the inner pipe, hot water flows from the
torus, and along the outer pipe towards it - water from the refrigerator. Teplo from hot water is transferred to cold, it heats up, boils and turns into steam.

Steam is supplied to the turbine blades, it starts to rotate and drives the shaft (rotor) 7 of the generator into rotation .

Having given off heat, the reactor water returns to the reactor again, The wa heats up and goes back to the steam generator. The ring along which it passes through is called the first circuit. The steam that unleashed the tour binu, goes to the refrigerator 9 . There it cools and turns into water. Water again enters the steam generator (outer pipe), and it turns into steam again. This is the second ring with water and steam on called the second circuit.

The fission chain reaction is controlled by control rods made of materials that strongly absorb neutrons. It can be boron steel.

When such rods are lowered into the reactor core, the tion slows down until it stops completely, and, conversely, when lifting the rods out of the zone reaction rate and reactor power increase.

To ensure the safety of the reactor, there are also emergency riot rods that fall within 1-2 seconds into the active zone and stop the chain reaction.

All nuclear reactors have special biological protection to protect the operating personnel from hazardous radio active radiation that cause ionization of cell molecules.

The coefficient of performance (COP) of NPP on slow thrones are usually 25-35%.

The world's first nuclear power plant was commissioned in Russia in 1954 in

Obninsk (near Moscow).

Russia's first fast neutron power plant was built in Shevchenko (for desalting sea water and generating electricity), its diagram is shown in Fig. 50.

According to the calculations of scientists, 1 kg of nuclear fuel with its full use use replaces 2,000 tons of coal.

NPP fuel is enriched uranium. In a nuclear reactor in during the course of work, plutonium accumulates, which can be fissioned under exposure to neutrons with the release of energy. As a result of the reaction

74

74

fission and transmutation products accumulate in nuclear fuel,

many of which are highly radioactive, and some have a periodhalf-life of tens, hundreds and even thousands and millions of years is a long

living radioactive waste to be disposed of in reliable burial grounds.

The infrastructure of the closed nuclear fuel cycle includes the following industries: 1) mining and enrichment of uranium ores and natural uranium production; 2) radiochemical processing spent nuclear fuel; 3) burial of radioactive waste moves.

Types of nuclear reactors and their fuels

Modern nuclear reactors use a small part of the energy contained in uranium atoms.

The fact is that natural uranium consists of two parts (isotope pov) - uranium-235 and uranium-238. The share of uranium-238 is 99.3%, and uranium-235 - only 0.7%. The uranium-235 atom splits into two fragments under the action of slow (thermal) neutrons. To increase the pro the duration of the operation of the reactor without reloading nuclear fuel, uranium ore is pre-enriched. The resulting content uranium-235 increases from 0.7% to 3-5%.

Subsequently, fast neutron reactors were created breeders. Their peculiarity is that in the process of fission of uranium nuclei 235 also involves uranium-238.

In our country, the construction of a nuclear power plant is based on hull reactors with water under pressure VVER (water-cooled power reactor) and boiling channel uranium-graphite reactors RBMK (high power boiling reactor).

The principle of operation of these nuclear reactors is the same - inside the rethe actor has fuel elements - fuel rods, which consist of a zirconium alloy metal tube filled a mixture of uranium-235 and uranium-238 (lower part of Fig. 50).

In a VVER reactor, all fuel elements are placed in a steel casing, filled with water that is in direct contact with the fuel rod mi and cools them. The heat of an atomic reactor heats water under high pressure, it becomes radioactive. Therefore this water goes to the intermediate steam generator, where the water of the second tour turns into steam, directed to the turbine.

75

75

The RBMK reactor is filled with graphite blocks, inside of which holes are made. They contain thin-walled pipes (working channels) made of zirconium, in which the fuel rods are installed. Through labor water circulates under pressure, it removes heat from the fuel rods and

it partially evaporates. It's channel

reactor, and VVER - pressure vessel. VVER received more widespread than RBMK.

The advantage of RBMK is the possibility replacement of fuel rods without shutting down the reactor.

Initial construction costs

NPP is 1.5-2 times higher than during construction

CHP, but the cost of electricity is lower in

1.3-1.7 times. The first nuclear reactor was

built under the leadership of I.V. Kurchatova

(1902-1960) (Fig. 51). The creation of a nuclear power plant is great achievements belong to the scientists A.P. AlekSandrov, N.A. Dollezhal (Fig. 52) and others.

Figure: 51. I. V. Kurchatov

The main feature of energy

fast neutron reactors is the opportunity to obtain only thermal and electrical energy, but also at the same time to produce new nuclear fuel. Osnew fuel in fast reactors neutrons is an artificial chemical chemical element plutonium-239 and "passive" ny "uranium-238.

Thermal energy in the reactor at fast new neutrons is obtained by fission plutonium nuclei, while part of the neutrons are captured (absorbed is) uranium-238 and it turns into plutonium-239. Newly formed plutonium is a nuclear fuel. In addition, in
the reactor produces an excess of new plumFigure: 52. N.A. Dollezhal tonium compared to fading, and it can be extracted from this reactor and sent for use in another reactor.

As a result of this process, it is possible to use almost all th uranium -238.

76

76

In 1972, the world's first nuclear power plant was put into operation in Shevchenko with the BN-350 fast neutron reactor with sodium coolant with a capacity of 350 MW.

A power unit was put into operation at the Beloyarsk NPP at a fast strong neutrons BN-600 with a capacity of 600 MW.

As a coolant and a reactor coolant at the BN, we use a liquid metal is produced - sodium, which in the subsequent circuit gives emits its heat to the water, turning it into steam, which goes into the steam turbinu; then the cycle of converting steam energy into electricity occurs walks in the usual way.

The sodium coolant should not combine with water anywhere, otherwise, an explosion will occur. Instead of sodium, it is designed in the future use helium or dissociating gas.

Here is the structure of electricity production in the largest countries of the world at the end of the 20th century.

table 2

Electricity production in some countries

Country	Electricity production, TWh			
	About- elec- electro-	Thermal electro-	Atomic electrotrostan-	Hydroother my stations
Total	in 13720	8592.0	2415.6	2516.7

the world Including: USA	3677.8	2518.7	720.8	353.1
China	1080.0	877.7		14.3
Japan	1012.1	601.2		304.6
Russia	847.2	577.4		109.0
Canada	570.7	118.1		93.0
Germany	555.3	361.5		161.6
France	513.1	43.1		401.2
India	435.1	367.5		8.4
United Kingdom 347.9		243.5		95.0

3.5.2. Renewable energy sources - RES

Renewable energy sources - solar energy, energy wind power, energy of rivers and streams, tides, waves, biomass energy sy (firewood, household and agricultural waste, animal waste innovation, poultry, forestry, woodworking and cellupaper industry), geothermal energy, and

77

77

dissipated thermal energy of air, water, oceans, seas, water emov.
All this diversity is reduced to three global types of sources nicknames:
-       the energy of the sun,
-       the heat of the Earth,
-       the energy of the orbital motion of the planets.

The widespread transition to renewable energy does not occur just because industry, machinery, equipment and life of people on Earth are cofocused mainly on fossil fuels. And also because that some types of RES are unstable and have a low density ness of energy.

The main advantages of RES compared to non-renewable energy sources (gas, oil, coal, etc.) are their
scoopability and environmental friendliness. Using them does not change energy balance of the Planet. In addition, RES play a significant role in

solving the three main problems facing human quality is energy, ecology, economics.

Consider the dynamics of the use of renewable energy sources in the world at the end of the XX century. Wind power (WU). Installed capacity worldwide: 1996 6172 MW; 2000 - 17824 MW; 2006 plan - 36,000 MW. Leadleading countries in this direction: Germany (6025 MW); USA

(2495 MW); Denmark (2364 MW); Spain (2538 MW); India (1214

MW); Russia (7.5 MW)

Geothermal energy (GeoPP). Installed capacity in world:

1970 - 678 MW; 2000 - 8000 MW. Leading countries: USA (2228 MW); Philippines (1908 MW) Italy (785 MW); Indonesia (589 MW); Russia (23 MW).

Solar energy (Solar power plant). Installed capacity in the world for 2000 - 260 MW. Leading countries: Japan - 80 MW; USA - 60 MW; Germany - 50 MW; Russia - 0.5 MW.

Biomass energy (BE).

The use of biomass energy goes in several directions niyam:

-        production of biogas and biomass in small plants (China,India - 6 million installations);

-        in large plants for the treatment of municipal wastewater(10,000 units) and on combined fermentation plants

78

78

urban and industrial waste water (more than 100 new the most recent installations);

-        at powerful combined installations (factories) for reprocessing waste products of agriculture, livestock

state, etc. (in Denmark there are 18 such installations out of 50

all over Europe).

Biogas is used in everyday life, in water heaters, steam boilers, diesel generators that produce electricity, etc.

Power plants are widespread (USA, Yesniya), which burn solid household production waste passages (solid waste) of cities, as well as power plants operating on biogas landfills (Italy, France).

Power plants begin to take root, in the furnaces of which wood, wood waste (Scandinavian countries) is used as with direct incineration of these wastes, and through their gasification with subsequent combustion of the resulting gas.

For several years, Russia has been training engineers specialists in renewable energy sources - in MPEI,

MSTU them. Bauman, Moscow State University, SPbSPU, universities of Yekaterinburg, NovosiBirsk, Khabarovsk, etc.

3.5.3. Promising sources of electricity

Thermonuclear reactor - TOKAMAK

In the process of studying nuclear reactions, it was found that it is advisable not only to divide the atomic nucleus of uranium or plutonium, but also to combine heavy hydrogen atoms (deuterium, tritium). When This produces a noble gas - helium. When merging (synthesizing) grathe yellow nuclei of hydrogen releases tremendous heat energy, exceeding the fission energy of an atomic nucleus per 1 kg of atoms mov.

In principle, it is possible to create reactors on hydrogen fuel, while as a waste of this reaction will be helium gas. Tawhich reactors are called thermonuclear and are called TOKAMAK (Fig. 53) - toroidal chamber with magnetic coilsmi. The thermonuclear process was discovered and carried out in a hydrogen bomb, but there it flows instantly and uncontrollable, and for use in the energy sector, it must proceed slowly and be controlled

79

79

edible (this process takes place at temperatures of about 100 million gram Dusov).

The TOKAMAK installation can be compared with a transformer, in which second, the secondary winding is made in the form of a closed hollow ring

tsa - torah. Filling such annular chamber heavy hydrogen nuclei - deuterium carried out in deep vacuum. When passing toka on the primary winding in the chamber breaks down in gas, the gas is ionized and heats up to a high temperature ratury.

Of course, this is where many problems, among them

Figure: 53. Installation TOKAMAK creation of devices that withstand temperature of many

million degrees. This can be done using a magnetic for which is able to keep plasma (ionized gas) from cotouching the walls of the device, protecting them from temperature round destruction. There are many more problems that arise. scientists around the world are trying to solve and are gradually solving them.

MHD - generator

It is a magnetohydrodynamic generator that can heat convert directly into electricity. At the same time, it is essential efficiency increases, since there are no intermediate transformations, but cheat, no "extra" losses.

In MHD generators, the gas, which is a plasma, consists of from a set of positively and negatively charged particles (electrons and ions), is passed between the magnets, which charged particles. Positive particles deviate into one side, and negative - to the other. These particles accumulate on two plates - electrodes and create a potential difference, i.e. create an electrical source of energy (Fig. 54).

Recall that Faraday's law of electromagnetic induction says that in a conductor moving in a magnetic field,

the electromotive force is applied. In this case, the conductor can be solid

80

80

smoke, liquid or gaseous. In our case, the plasma is z-shaped conductor.

Science field, studying interaction between magician field and current co-conducting liquid bones and gases, oncalled magnetoguide birthmarks. Thereforemu generators, working melting on plasma nominal conductor received the name magician nitrodynamic generators - MHD.

Figure: 54. MHD - generator

But! There are many problems here that need to be addressed. joint efforts of physicists, power engineers, materials scientists, etc. It is difficult to convert large masses of gas into plasma. This requires high temperature and high-energy fuel. At high temperature, it is difficult to preserve the materials from which the generator is built rator. There are other similar difficulties.

Ionospheric MHD generator

Life on Earth in all its diversity is directly related on with the activity of the sun. Billions of years, giant streams of solcertain energies rush to the Earth. For several centuries now, solar technology is developing, dealing with the use of solar new energy.

Not so long ago, a tempting prospect arose: to find a way to use of that part of the solar energy that does not reach our necks of the planet, scattered in space.

1.   Physical research in space showed that due to theabsorption of X-ray and ultraviolet radiation of the Sun in

ionosphere of the Earth is continuously generated (generated) of electricity tric current. It passes (flows) around the Earth in the western direction near the equator at a distance of several earth radii

from the surface of the Earth. Its energy in the form of Joule heat dissipates is found in the surrounding area.

81

81

2.   The earth has a magnetic field, and the plasma of the ionosphere crosseslines of force of this field and therefore in plasma, as in a conductor, is the electromotive force.

3.   In the shadow part, where the Sun does not currently illuminate the Earththe ionosphere is cooled, the concentration of its particles decreases due to recombination - the formation of electrons and ions neutral atoms, and from atoms - molecules. Through the plasma of the ionosphere a circular electric current flows, interacting with the magmagnetic field of the Earth, while the separation of charges occurs: part positive ions are concentrated on the inner shell and

nospheres, and some of the electrons - on the outer. As a result, on the surface the earth creates a negative charge.

Electrically conductive Earth and ionosphere separated from each other dielectric - a layer of non-conductive air, represent is a giant spherical capacitor. With a layer thickness spirit of about 100 km constantly generated potential difference fishing between the Earth and the ionosphere reaches about 100 million volts.

Since there is a circular electric current in the ionosphere, then we can conclude that at its various points there is always is the potential difference. It is equal to the product of tension fields at the distance between these points. To use this the potential difference for doing useful work on Earth is not

it is necessary to have only two conductors connecting the sections of the ionspheres with corresponding points on the surface of the Earth, and useful load connected to these conductors.

The main problem for this way of using energy The sun is the creation of conductors that connect useful load with the ionosphere.

Hydrogen energy

The term "hydrogen energy" assumes widespread use the formation of hydrogen in energy systems and other sectors near the best future.

In order to appreciate the advantages of hydrogen over with other types of fuels, we present the following data.

82

82

When burning one kilogram of the following fuels heat is generated:
20,000 kJ of heat - dry wood;
13000 —— "—— - brown coal;
25000 —— "—— - anthracite;
42000 —— "—— - oil and oil products;
45000 —— "—— - natural gas; 120,000 —— "—— - hydrogen.

Consider as an example the use of hydrogen in the top shower elements for generating electrical energy. In contrast from thermal power plants, which chemical energy fuel first converted into heat, and only then into electricity, into fuel element, there is a direct transformation of chemical energy into electrical energy. The main difficulty is that both the fuel and the oxidizer must first be converted into ius. The fuel in these cells is hydrogen, and the oxidizing agent is more oxygen, or air.

The schematic diagram includes a hydrogen anode, an oxygen cathode and electrolyte, conducting ions.

For example, in a fuel cell (with alkaline electrolyte), which serves as a source of energy for spacecraft, dissociation and ionization of molecular water occurs at the anode kind:

$$H_2 = 2H^+ + 2e$$

An alkali solution is usually used as an electrolyte
KOH. Hydrogen ions + $^+$ H under the influence of the potential difference between du anode and cathode diffuse through the electrolyte layer to the cathode.
Electrons formed at the anode - e when the external electrical circuits flow to the cathode, doing useful work. A reaction occurs at the cathode:

$$4H + 4e + O_2 = 2H_2O^+$$

The only reaction product in this process is
water vapor or just water.
In fuel cells, ionization occurs at moderate temperatures in the presence of catalysts including metals platinum group. The efficiency of such elements is from 40 to 70 percent.

83

83

Many studies have shown that under the thermodynamic effect efficiency of steam turbine and steam-gas hydrogen-fired installations with a capacity of 1-10 MW are close to fuel cells, and specific power exceed them.

3.6. Energy development in Russia

Power plants were created in Russia at the end of the 19th and the beginning of the 20th centuries, however, the rapid growth of the most important component of the development of bogo production - electric power and heat power - began in the twenties of the twentieth century, after the adoption at the suggestion of V.I.
Lenin's plan GOELRO (State Electrification of Russia).
Major scientists and experts took part in drawing up the GOELRO plan. engineers, experienced electrical engineers of the country. The preparation of the GOELRO plan is preceded by la great scientific and organizational work. Based on numerous specially designed prepared materials were developed

detailed scientific reviews including self reports on the state of technology, industries industry, agriculture,

construction and transport, territorial location of industrial centers,

Figure: 55. G.M. individual large enterprises and their
Krzhizhanovsky technical technical equipment; detailed characterization

teristics of energy, labor, raw materials

output and monetary resources of the country's economic regions.

Supervised drafting plan
GOELRO G.M. Krzhizhanovsky (1872-1959)
(Fig. 55) took part in the development of the plan such eminent specialists as A.V. Winter (fig. 56), G.O. Graftio (fig. 57), B.E. Vedeneev, A.V. Wolfe et al.

GOELRO plan was adopted in December 1920, and further energy production gram of the USSR for the long term
was adopted in April 1983. This approach
Figure: 56 A.V. Winter
provided scientifically sound and lanced development of the main directions of the country's development.

84

84

The GOELRO plan took into account the dependence of productivity labor from the level of development of technological equipment, providing electrification of production processes; were hall-
wives theoretical principles of formation, confirmed by experience the development of the world economy.

The electrification of the country in the GOELRO plan acted as an economic the most favorable beginning of the national economy, the most important factor of technical progress, one of the means of solving the primary rare social problems of Russia.

GOELRO plan - the country's energy development program was completed by 1931 - the minimum period for the implementation of the plan. TO 1935 - the maximum period for the implementation of the plan - the program was significantly overfulfilled in terms of the number of erected power plants

tions, and by their qualitative and quantitative characteristics.

This allowed the USSR to take second place in Europe by 1941.

(after Germany) and third in the world (after the USA and Germany) in terms of electricity generation

Until 1913, Russia stood at 15 place in the world.

The plan paid much attention to the problem of using local energy resources (peat, river water, local coal and etc.) for the production of electrical energy.

In 1922, the power plant was launched "Utkina Zavod" - the first peat electric station in Petrograd, in 1925 - Shaturskaya peat power plants. In 1925, KaShirskaya power plant began to develop new technology of burning the Moscow region coal in the form of dust.

It was put into operation in 1924 the first heat pipeline from LGES-3 (CHP named after L.L. Ginter) - this marked the beginning of the development district heating in Russia. In 1926, the power was put into operation

Figure: 57. G.O. Graftio naya Volkhovskaya hydroelectric power station, energy

the power of which, along a 110 kV power transmission line,

with a length of 130 km was delivered to Leningrad. Supervised construction Volkhovskaya HPP Genrikh Osipovich Graftio (1869-1949) (Fig. 57).

85

85

In 1921, energy associations of the state were created military power plants in Moscow - MOGES, and in Petrograd - "Petrotok "(then renamed to Lenenergo).

The first 110 kV power transmission line Kashirskaya GRES - Moscow was commissioned in 1921.

During the implementation of the GOELRO plan, in practice, basic principles for the development of the electric power industry, namely:

-        construction of large power plants (thermal and hydrostations), allowing to provide power supply to entire regions onov;

-        use of local energy resources;

-        construction of high-voltage transmission lines;

-        equal distribution of energy resources across the country, etc.

Electrical engineering is the main branch of electrical technical industry manufacturing generators for power the chemical industry and electric motors for various branches of the national economy.

Before the revolution of 1917 in Russia there were only four mechanical engineering plant factories: the plant of the Siemens and Schuckert concern in St. Petersburg, plant of the electrotechnical trust "Westinghouse" (plant "Dynamo") in Moscow, the plant of the concern "General Electricity Company" in Riga and the Volta plant in Revel. These factories were assembly workshops working on foreign semi-finished products and materials riyalah. The scientific and technical base was located abroad.

The electrical industry of Russia fell into high dependence on foreign capital. Has become bitter right fork, when the world-famous Russian electrical engineers are the first the discoverers, finding no use for their forces at home, were looking for applications in a foreign land: P.N. Yablochkov, A.N. Lodygin, M.O. Shares in-Dobrovolsky, etc.

At the end of the 19th century, 75% of all electrical products were supplied were brought to Russia from Germany, incandescent lamps were brought from America and such a listing could be continued.

V.N. Chikolev united young Russian electrical engineers during circle of the magazine "Electricity", on the s of which began to appear to publish accusatory articles against the Russian government. This "

A mighty handful of "electrical engineers began to fight against the dominance of foreign strange capitals, foreign specialists and technology for the fact that-

86

86

would be government and public orders for electrical Russian companies made their products.

Gradually, the domestic electrical industry broke through your way. The first-borns of the Soviet electrical engineering were factories "Electrosila", "Dynamo", KHEMZ (Kharkov Electrotire building plant).

The plant "Electrosila" focused on the development and production management of turbo- and hydrogenerators, powerful electric motors standing current for rolling mills and ships, asynchronous machines.

The Dynamo plant specialized in traction and crane electric tro-equipment, as well as electrical equipment of hydraulic structures. A series of specialized machines were created.

The first 500 kW turbo generator was built in 1924 at

"Electrosile", and in 1937 a turbine generator for 100 thousand kW with air cooled. In 1946, a turbine generator with hydrogen cooling. In the post-war years were built

Novosibirsk (NTGZ) and Lysvensky (LTGZ) turbine generators plants.

As a result of the use of new cooling systems and new constructural, magnetic and insulating materials with power speakers grew in the following steps: 100, 200, 320, 500, 800 and 1200thousand kW.

At the same time, hydroelectric power engineering was developing. In 1923, waters "Electrosila" built a hydroelectric generator for the firstborn of the Soviet hydropower - Volkhovskaya HPP, then for Uglichskaya and Rybinhydroelectric power station, Volgograd hydroelectric power station, etc.

During the years of the Great Patriotic War (1941-1945) there was a loss but, destroyed, plundered 60 large power plants, removed from building 10 thousand km of power lines.

Exported to Germany: 1,400 steam boilers, 1,400 steam and hydraulic roturbines, 11,300 large power generators and hundreds of thousands of trodmotors.

Already in 1947, thanks to the tremendous work of the Soviet people, Russia comes out on the 1st place in the production of electricity in the EU rope, and in 1958 - on the 2nd place in the world.

In 1954, the first nuclear power plant in the world was built in the Soviet Union for 5 MW.

In 1958, Russian science and technology created a turbogenerator torus for 200 MW, in 1959 - for 220-300 MW and more.

87

87

The district heating of plants, factories, city dov, villages, aimed at industrial and communal household needs.

The total capacity of cogeneration turbines is over 36 % of the total capacity of thermal power plants. Wide central named district heating saves many million tons of conventional fuel per year.

After 1920, intensive construction of hydroelectric stations; the largest of them:

Volzhskaya - 230 MW;

Gorkovskaya - 400 MW;

Rybinskaya (on the Volga) - 330 MW;

Kamskaya - 504 MW;

Tsimlyanskaya (Don) - 166 MW;

Dneprovskaya - 648 MW;

Irkutsk (Angara) - 660 MW; Sayano-Shushenskaya - 6400 MW.

For comparison: the largest in the United States, Grand Coulee (R. Columbia) has a capacity of 1,974 MW.

The total length of power grids in the Soviet Union is only voltage of 35-400 kV in 1958 amounted to 100,000 km.

At the same time, the production of oil-filled cables for 220 kV.

By 1990, the production potential of electricity in Russia amounted to: ~ 700 power plants (of which 70% CHP and KES, 20% HPP and 10% NPP).

The electric power industry had 2.5 million km of power lines of all voltage classes (including 150 thousand km with voltage 200-1 150 kV); 90% of this potential is concentrated in a single energy system - UES, which has a centralized and operational dispatch control.

The transmission of electrical energy over long distances posed the country has the following tasks:

-        voltage increase: 110, 220, 330, 500, 750 kV and above;

-        transition to transmission by direct current due to high agemelting of capacitive currents at high alternating voltage (> 220 kV);

-        creation of high-voltage rectifiers and inverters;

88

88

-        increasing the dielectric strength of insulation and insulators;

-        creation of high-voltage switches and lightning protection;

-        creation of special cables - underground and submarine;

-        creation of oil-filled cables for 110-500 kV;

-        replacement of copper wires for overhead lines with aluminum onesand steel-aluminum.

3.7. Energy systems

Combination of power plants carrying out parallel work on a common network is called an energy system. First a power system was created in Switzerland in 1892 (the firm "Airliecon "), in Russia they were created in 1902.

Modern energy systems consist of:

1) power plants generating electricity and heatpilaf energy;

2) transformer substations serving to transformelectrical voltage of individual elements of the system to economical a reasonable level in accordance with the range of supply of electricity and the amount of transmitted power;

3) high voltage power lines, through which electrictric energy is transmitted over long distances to the centers of loads of certain areas;

4) distribution networks of various voltages supplyingenergy directly to consumers;

5) consumer installations, consisting of motors, electricovens, lamps, etc.

Such an energy system contains tens of thousands of times personal elements working together at different voltages at different frequencies.

For example, the Lenenergo high-voltage network (founded in 1926) includes 159 substations for 330 kV, 220 kV, 110 kV and 35 kV. Its total length is 3,311 km. The substations have 475 transformers with a total capacity of over 19,000 MVA.

Power plants included in the power system, power lines transmissions, substations and heating networks are connected into one whole mode and continuity of the production process and distribution division of electrical and thermal energy.

The creation of power systems has a large national economic value. When working together on a common electrical network of a number of electrical

89

89

the total system load is appropriately distributed between them, a more economical use of equipment is achieved individual power plants and energy resources (fuel va, water), as well as reduce electricity losses in networks and, hence, the cost of electricity is reduced. Besides, the overall reliability of power supply increases significantly consumers.

The creation of large unified energy systems allows regulating electricity supply in accordance with different time zones countries or different countries and with different peak hours of the day loads (covering maximum loads).

In large power systems, there is a main dispatcher

Chersky center, which receives a continuous stream of information about capacity, loads and energy distribution throughout the region (part of the country or the whole country).

Such a system can include hundreds of power plants, substations, transformers and thousands of kilometers of high-voltage transmission lines.

The dispatch center receives a large number of necessary information, maintains communication with other systems and is the connecting center of the united systems. From it, all switchings in the high-voltage network are found. Dispatching the center is equipped with automation and telecontrol systems.

Consider the history of the creation and development of energy systems.

The growth of energy is associated with the formation of a unified energy systems. UES is the most complex, artificially created system, functional rationing and management of which is a complex scientific technical and economic problem. This is due to the centralization of power supply and concentration of power

sty, complication of the structure of electrical connections, an increase in dependence of the electric power industry on the components of the fuel the energy complex of individual large regions, from the world the fuel and energy situation, the growing influence of the level on reliability of power supply for the functioning of the country's economy generally.

By 1983, there were three most powerful energy systems in the world.

Topics (power interconnection):

-       North America (USA and Canada);

-       Western European (Austria, Belgium, Italy, Luxembourg,

90
90

Netherlands, Germany, France, Switzerland);
-	Unified Energy System (UES) (USSR, Bulgaria, Hungaryria, Poland, Romania, Czechoslovakia, East Germany).

The basis for the dispatch control of such systems are information and computing systems (ICS), including computers, means of automation, telemechanics, etc.

Creation of unified power systems, connection in parallel bot of the energy systems of neighboring countries, the formation of powerful interstate free energy networks characterize the development of the world power engineering (Tables 3, 4).

By the beginning of the 80s of the twentieth century. almost 90% of power plant capacity of the world was concentrated in the established energy systems.

Table 3

Unit power and voltage levels of the main power plants in some countries in the 80s. XX century

Country	Power most power plants, MW			Power most largelarge unit, MW			Highest lev voltage, kV	
	NPP	HPP	TPP	NPP	HPP	TPP	1980	2000
the USSR	4000	6000	3800	1000	600 1200		750	115
USA	3200	6200	3200	1100	700 1300		765	120
Japan	4700	1280	4400	1175	320	1000	500	100
France			-	1300	-	700 4200 800	400	400
Sweden		-		one hundred800 - 330 1600			400	800

Electricity production data

Table 4

Country	Territory, thousand sq. km	Population, million people	Electronic productic energy, billion kWh for 1980
the USSR	22402	264.5	1294
USA	9363	219.9	2475
Japan	372	115.3	620

Country	Territory, thousand sq. km	Population, million people 1980	Electronic production energy, billion kWh
Canada	9976	23.5	389
China	9597	975	301
England	244	55.9	285
Italy	301	56.7	183
France	551	53.4	255
Brazil	8512	115.4	137
Poland	313	35	122
India	3288	634.2	112

Continuation of table 4

Country	Territory, thousand sq. km	Population, million people 1980	Electronic production energy, billion kWh
Spain	505	36.2	110
Sweden	450	8.3	95
Norway	324	4.1	82.6

3.7.1. Development of power systems in the USSR

The first attempt to create a power system in Russia was adopted in 1902 on the basis of two power plants in Baku under the leadership by engineer R.E. Klasson.

The process of combining power plants for parallel operation and the formation of larger energy systems began in our country in the 20s. By the end of 1935, 6 power systems were already in operation. Prothe length of 110 kV electrical networks was 2 thousand km, in 1933 the first power transmission line (PTL) was built voltage of 220 kV.

In 1942, the foundations of the unified power system were laid Ural.

Many power systems operated in our country, the main ones being fox: Ural, Central, South, North-West. Most-

in the energy systems of the European territory were combined into a Unified power system in 1962, and in 1972 there was a merger with this system of energy systems of Western and Eastern Siberia and education the establishment of the Unified system of the country.

At the end of the 50s. a transmission line with a voltage of 500 kV.

By 1970, the creation of the Unified Energy System of the EuroPeysky part of the country. At the beginning of 1978, the formation of the UES by connecting the energy systems of Siberia.

In 1981, 96 regional power systems operated in the country. Parallel The efficiently working district power systems had a common center operational dispatch control.

Structure of UES capacities at the beginning of 1982: HPP - 19.3%, NPP - 6.5%, thermal power plants - 74.2%. The main electrical networks that form UES - networks with a voltage of 330, 500 and 750 kV.

The 500 kV network is the main backbone of the UES, providing reception and output of power of the largest power plants

92

92

tion and exchange of power between power systems. Length such networks amounted to 25.4 thousand km.

The introduction of 750 kV voltage began in 1974-75. and by 1981 the length of 750 kW transmission lines exceeded 3 thousand km. 90e years, the UES has high-voltage transmission networks for 110, 220, 500, 750 and 1150 kV AC and 1500 kV DC (Table 5).

Table 5 Length of AC power lines with voltage
220-800 kV at the end of 1980

Voltage, kV	Length of lines, thousand km			
	the USSR	USA	France	Japan
220 ... 300	92.8	105	24.4	21
315 ... 345	24.3	55	-	-
380 ... 460	0.6 - 10.1 480 ... 550 24.9 29 - 3.2			
750 ... 800	2.9	2.6	-	-

Out of 93 energy systems of the country as part of the EEC for 1980, worked 67 power systems, providing power supply to the national economy of the European territory of the country, Transcaucasia, Urals, North loyal to Kazakhstan, Western Siberia.

Electricity was exported from the Unified Energy System.

power supply to the parallel power systems of Bulgaria, Hungary, Czechoslovakia, Poland, East Germany, Finland, etc.

Before the collapse of the USSR, there was an interstate dynasty "Mir", which included the power systems of socialist countries, Finland, Norway, Iran and Mongolia.

The main fuel and energy resources in European regions of the country are - atomic fuel, in the Urals - Kuznetsk and Ekibastuz coal, in Siberia - Kansk-Achinsk coal and water resources sy, in Transbaikalia, Yakutia and the Far East - local coal, gas and hydro resources.

3.7.2. Power supply

Electricity supply is called the provision of consumers with electricity tric energy. Consumers mean enterprises

services, organizations, workshops, construction sites, residential complexes, etc., at which receivers consume electrical energy. Receiving Kami of electricity are devices in which there is

transformation of electricity into other types of energy: mechanical

93

93

sky - electric motors; thermal - electric furnaces, welding aggregates; chemical - electrolysis baths; light - light

nicknames; information - devices for receiving, transmitting, controlling and information processing (radio, telegraph, computers, television sets and etc.).

The first installations supplying enterprises with electricity appeared in 1842, when the Englishman D. Woolrich first connected DC electric generator with steam engine. By-

after that, such installations began to be introduced into factories and other enterprises. There was a need to build an electrical area nye and central power plants.

In 1879 T. Edison developed a program with which, in essence, woo and the science of electricity began. It includes the following main points:

- develop powerful generators adapted toparallel connected electrical receivers;

- develop rational schemes for the distribution of electricaltrienergy;

- develop a reliable design of conductors and wayssob their laying;

- develop protection of the power supply system from currentsshort

circuit;
- make simple and safe switches;
- find ways to regulate the voltage of generators;
- manufacture metering devices for electricity supplied;
- develop a system for standardization of parameters and sizeditch of lamps, apparatus, wiring parts and other devices.

Note that the first, simplest and most basic protection against short circuits served (and is) the floating cue fuse.

With the development of electricity consumption and power plants, there was the need to assess the effectiveness and maximum transmission range chi electricity.

The first formulas for determining the efficiency of power transmission were given in 1877 Frenchman E. Mascar. Mathematical calculations of the transmission of electrical long-distance power generation was first developed by Russian Skim scientist D.A. Lachinov in 1880; research in this direction was led by the Frenchman Despres.

94

94

Of the greatest importance was the invention of M.O. Topping up Dobrovolsky three-phase AC system in 1888.

In 1892 Dolivo-Dobrovolsky determined the dependence on power loss in the transformer from the nature of the load, applying decomposition of currents into active and reactive components

U

I a

I C   I L

The concept of "power factor", related to the number of important most important in power supply, was introduced in 1884 by an Italian Ferraris.

To increase the power factor M.O. Topping up

Dobrovolsky was the first in the world to install a synchronous compensator.

Reactive power balance and solution to the boost problem factor of reactive power using capacitor banks

tarai was proposed in the 1890s by the Frenchman Paul Bouchero. After creating cheap waxed capacitors (approximately in 1915) the problem of reactive power compensation was reshena.

In 1884, the Englishman Parsons created a stepped jet turbine, connected it to the generator and received a turbine generator - important The most important unit of thermal stations.

3.7.3. Development of electronic computers

Development of the electric power industry, creation of large power systems demanded the widespread introduction of automation.

Let's trace some historical milestones of this implementation on the basis of new monitoring and control devices, computers, computers (computer - calculator).

As early as 1500 years ago, abacus began to be used.

In 1642 Blaise Pascal (1623-1662) invented a device for mechanical numerical addition and subtraction.

In 1673 G.V. Leibniz constructed an adding machine, calculated ny on four arithmetic operations.

95

95

In the first half of the XIX century. Englishman Ch. Babbage developed the idea of a computing device that works without human intervention ka. In 1941, the German Konrad Zuse built an analytical engine. In 1943, the American Howard Aiken developed and built a computer casting machine based on mechanical relays "MAPK-1". In 1943 year D. Mauchly and P. Eckort built an electronic computer

new machine based on electronic tubes (instead of relays) - a computer.The

first computer was developed by mathematician John Ney. man, and built in

1949 by Maurice Wilkes.

It should be noted that the Mauchly and Eckort computers had 18 thousand electronic lamps, 15 thousand relays and consumed 150 kW of electricity gii. The program of this machine was typed manually on the commutation discs and switches.

The first generation of computers worked on tube circuits.

The second generation of computers already used semiconductor transistor circuits. Performance and reliability have increased, less energy consumption. Number of operations - 1 million give me a sec.

For the third-generation computer, small integral schemes (MIS). They contain electronic devices and elements for special new technologies are created already in the substance itself - the crystal. Yet the dimensions of the computer and the power consumption have decreased more, performance has increased; the machine began to process not only digital information, but also alphabetical.

Computers of the fourth generation are already built on large integral ny schemes (LSI). The number of elements per 1 cm 2 $_{of\ the}$ crystal is adjusted to 100 000.

Already from the beginning of the 70s. XX century microcomputers are being introduced into all spheres ry of human activity.

The fifth-generation computer performs not only numerical calculations, but carries out the processing of semantic information with the implementation analysis and output operations. Manufactured on LSIs, VLSIs and microcomputer.

In Russia, the first small computer "MESM" was built in 1951 under the leadership of S.A. Lebedev. She had parallel processing of codes and consumed a power of 25 kW. IN

1953 appeared "BESM", which had a much faster

act. In the same 1953, a computer was created - "Strela" under the leadership

96

96

by Yu.A. Bazilevsky. In 1954 a computer was built "Ural" under the leadership of B.I. Rameeva.

After the creation of these machines and as experience gained, steel machines of the next generations, up to the sixth, were created.

The use of electronic computers, computer ditch made it possible to solve complex problems in the field of transitional and new modes of electrical machines, distribution of electrical energy in systems, etc.

3.7.4. Overhead power lines (VL)

The first 110 kV transmission line was built at

U-shaped wooden supports. But the tree rots quickly, you need it treat with an antiseptic (since 1932 - the service life is 30-

40 years), while the distance between the supports is 50 ... 100 meters.

The first metal supports in the USSR were built in 1925.

(Shatura-Moscow), the distance between them is 200 m.

Then they began to build the supports of the spatial structure -

Verkhne-Svirskaya HPP - Leningrad (220 kV). Later applied portal structure of supports - the span increased to 350 m (higher wires, a grounded steel cable was suspended to ensure lightning-proof transmission lines). Then they began to use iron ton supports.

In 1955, a line was built in Italy through the Messina liv. Its length between the supports is 3,650 m. The line in Norway, between the rocky shores of the fjord has a long

distance between supports 4800m. The usual phenomena spans from 50 up to 100, 150, 350 m.

Currently me continues withimprovement of the line support structures gears (fig. 58).

For voltage

1150 kV tower height

Figure: 58. Supports of high voltage lines should be 45 m, so that the height of the wire

from the ground was 17-23 m, and the distance between the phases was 23 m (Fig. 58). Op-

97

97

the minimum wire cross-section is 300 mm 2 , the phase consists of 8 steel-aluminum wires located at the tops of the correct polygon.

Back in the 19th century. M.O. Dolivo-Dobrovolsky said that the future transmission of high voltage electricity for direct current.

The use of direct current transmission systems makes it possible ability to increase the dynamic and static stability of the joint energy systems. The following are the benefits of whether DC transmission:

•   transportation of electricity is carried out in two wayswater instead of three (with three-phase current) - consumption is reduced non-ferrous metal by 1/3;

•   the supports are much lighter, since the weight of the wires is reduced;
•   loss of magnetization reversal of wires disappears, asstanding current does not change direction;
•   power transmission can be adjusted by hardwareestates;
•   generators can operate in asynchronous mode when connectedtransmission lines of energy systems.
Disadvantages of lines overgiving direct current:
-          the presence of two substationstions with transducers ka, (Fig. 59), significantly increases the cost of electricity transmission;
-          power transmission actingvines like a transit, without

Figure: 59. Transformative high-voltage substation (800 kW) direct current ~ 85%.
bora of electricity on the way. Economically beneficial
Overhead line for direct current from 1
500 kV and above, line efficiency at

VL direct current extra-high voltage, except transmission large amount of energy over long distances, at optimal combined with AC overhead lines play an important role in the UPS.
To address issues of economical energy transmission, appropriately address the issue of reducing line resistance transmission.

98

98

You can significantly reduce the resistance of conductors by deep cooling them. Such conductors are called guconductors.
In 1911 the Dutch physicist Kamerling-Onnes discovered the phenomenon superconductivity. The idea of superconducting lines was adopted since great enthusiasm. But getting low temperatures is not very good just. Liquid helium gives $t° = 4.2$ K, and liquid hydrogen - $t° = 20$ K.
For more than four decades, superconducting conductors have not could be used in practical designs.
In the mid 50s of the twentieth century. alloys of niobium with tin were

discovered vom, which went into a superconducting state at a temperature round twenty kelvin (20 K); ceramic compositions - T cr = 35 and 78 K.

Studies have shown that superconductors can be successfully apply for direct current transmission (with alternating - losses more).

The most important condition for the operation of power lines is their reliability. A whole system of power transmission line protection has been developed.

Lightning protection is used against atmospheric discharges in the form steel cables located above the wires, lightning discharge nicks, arc suppression coils, arresters - surge arresters semiconductor, etc.

In order to avoid system accidents that may occur when damage to supports, wires, insulators of transmission lines, applies

Xia relay protection - high-speed relays and devices. Wide APV is used (automatic reclosing devices niya) and ATS (automatic transfer switches). Out of 100 cases of emergency shutdowns of consumers and their reverse after a few seconds (using automatic machines) in 80 cases, a normal power supply is established without disrupting production management process.

3.7.5. Heat supply and district heating

Many industrial enterprises require for their technology logical process a large amount of thermal energy in the form steam or hot water, e.g. chemical plants, textile factories, etc. In addition, a large amount of heat energy required in cities and businesses for heating purposes.

99

99

Before the development of a wide network of power plants, heat supply was was produced from boiler plants erected at the enterprise y and separate houses, and this required a high-calorie fuel fuel.

It was more economically feasible to switch to a comprehensive mu district heating, heat and electricity energy from combined heat and power plants - CHP.

The beginning of district heating in our country is considered to be 1924, when the first cogeneration plant was put into operation at one of the Lenin city power plants.

Since the transmission of heat energy over long distances is it is difficult to fill up due to large heat losses in pipelines, then CHP are usually built near cities and large enterprises.

With the combined generation of heat and electricity energy at the CHPP saves a large amount of fuel, and besides th, the efficiency of such stations is much higher than that of condensing stations, in which only electricity is being generated.

Specific fuel consumption - 338 g per 1 kWh of electricity.

This flow rate is determined by the power of the plant units and the parameters steam rami (Table 6).

The increase in steam parameters is associated with the development of production structural materials, their strength characteristics, onreliability of drums of steam boilers, flow parts of turbines, pipes sidelines.

In addition, an increase in steam parameters presents special requirements for the chemical purity of feed water, its high deaeration (i.e. removing oxygen from it).

Table 6 Fuel consumption depending on steam parameters and unit capacity

Power Steam parameters Ud. consumption, unit, thousand kW Pressure, atm

Temperature, ° C g / kWh

24 - 25	30 - 35	400 - 425	520
300	240	560	320

Heating or centralized heat supply management and residential premises was developed in the USSR in 1924, the first originally in Leningrad (it was designed by engineers Dmitriev and Gunther).

100

one hundred

The provision of electricity and heat in the modern world is is the basis for the well-being and development of society and any state gifts. The stability of existence and the ability to work any industries of the population are determined primarily safety and reliability of electric heat power systems. AND

this is confirmed by the periodically occurring

personal regions of the world disrupting their work, when whole areas or millions of cities are left without electricity, people are stuck

they spend hours on the subway, elevators, heating in houses freezes, deteriorates tons of products, animals die, etc.

It is absolutely clear that it is necessary to reduce as much as possible the cost of electricity and heat energy, and its harmful action on the natural environment and the person himself.

3.7.6. Energy thresholds

Academician Gleb Maksimilianovich Krzhizhanovsky formulated shaft the concept of "energy thresholds" - periods when, as a result those qualitative improvement of the energy base occurs leaps in labor productivity growth, especially in labor-intensive processes - physical and mental.

Among such rapids: a water wheel and a windmill, paroving machine, industrial use of electricity, electric and radio electronics, computer technology, etc.

Such energy thresholds so strongly influence development productive forces, which are also determined by the characteristic stages of the mother the spiritual culture of a person as a whole.

History knows several energy thresholds.

1.   The first threshold is to create a water wheel that is effectivereplaced the muscular strength of humans and animals - belongs to the 3rd thousand BC e.

Then other sources of energy were used: wind - sail ny boats, windmills; smelting metals with energy fossil fuel, etc.

2.   The second threshold - the first half of the 18th century - the transition from thefrom industrial production to machine production. Energy the basis for this was the universal steam engine and the development of mechanical Ki and heat engineering. The steam engine was fundamentally new an energy engine that converts chemical energy fossil fuel into thermal energy (creation of water vapor

101

101

pa), and after that into mechanical - this required the development of scientific Noah thought in the field of mechanics and thermodynamics.

3.   The third energy threshold arose and was implemented on the ruthe beginning of the 19th-20th centuries, when a large concentration of production was required production, which could no longer be provided by

low-power steam tire.

The development of the electric power industry has served to create new primary ny engines - steam and hydraulic turbines. Their combination with electric generators, transformers and power lines transmissions, receivers of electricity - electric motors, blowing devices, electric furnaces, etc. created principle pialally new energy base and ensured a relative independence of the location of power generation sources from the centers of its consumption.

4. The fourth energy stage (threshold), in timecoinciding with the third, was the creation of a fundamentally new engine - internal combustion engine. This made it possible to create transport vehicles - cars, locomotives, etc. They work

are due to the direct conversion of the chemical energy of the fuel into mechanical energy.

5. In the second half of the XX century. humanity is moving to a new energy threshold. It is characterized by a combination of a number of energy development boards: using a qualitatively new energy resource - nuclear fuel. This energetic

the threshold is combined with the massive development of electronics, changing many industries that require the application of mental work, create the use of computers, robots, industrial automation, etc. Self-test questions

1. When did the first power plants in the world appear, in Russia?

2. What types of power plants exist on non-renewableenergy sources?

3. What are the types of power plants on renewableenergy sources?

4. What promising devices are being developed for semiof electric energy?

5. How did the electrification of Russia develop until 1917? Where and what large power plants were built, who was in charge construction?

102

102

6. What is the GOELRO plan, its authors?

7. What power plants and power lines were builtare we on the GOELRO plan?

8. What place in the world did Russia occupy in the generation of

electricityenergy in 1913, what by 1941?

9. What losses in the electric power industry of Russia took place during thethe name of the Great Patriotic War?

10. When did the centralized heating at the basethermal power plants in Russia?

11. What energy thresholds do you know and how do they characterizeriot?

4. ELECTRICITY AND ITS EFFECT ON ENVIRONMENT

The process of generating electricity and heat on electricity stations is accompanied by significant interaction with external environment.

Earth is a unique celestial body in the solar system. Her post the surface and its environment have several areas: lithosphere - solid surface ;

hydrosphere - the watery part of the Earth ; atmosphere - air part ; biosphere - biological part ; noosphere - intelligent sphere, sphere

mind (sphere of interaction Naturedy and man, where a reasonable man activity is the main decisive factor).

Outstanding Russian scientist Vla-

Dimir Ivanovich Vernadsky (18631945) (Fig. 60), developed the doctrine of biosphere, and later on the noosphere, coaccording to which all living organisms,

Figure: 60. V.I. Vernadsky including humans and their habitat organically linked and interact with each other, forming a chain Lost dynamic system.

103

103

The biosphere on Earth arose under the influence of solar energy and long-term biochemical processes. Man as an element of the biosphere ry, appeared relatively recently, in one of the later stages of its evolution lucia. In the entire history of mankind, about 20 thousand now generations of people.

Before there was cattle breeding and farming and there was the transition from gathering to production, a person lived in full harmony with Nature. Then the person began to actively take root in native sphere and to remake nature more and more intensively.

Until the twentieth century. human activity was more or less local character, and in the twentieth century. it covers the entire biosphere and is already globally ball character.

Jacques Yves Cousteau wrote:

"Previously, nature frightened man, but now man frightened nature. "

In his works of the last period V.I. Vernadsky pointed to the need for the transition to the noosphere - the sphere of reason, when the economy the human activity will be exactly commensurate with its influence I eat on all spheres of the Earth, and especially on living matter.

Nature is a living organism, where the interdependence of all processes are self-regulating. Like any organism, it adapts it is done by adapting to changing conditions. However, obprogress has now reached such a level that mentality and energy have a significant impact on biosphere, lithosphere, hydrosphere and atmosphere due to commensurable capacity of processes in man-made systems and globall processes in Nature .

Let's look at the value of the capacities of the processes occurring in nature and in man-made installations:

Radiation from the Sun to the Earth - 10 17 W;

All rivers and waterfalls on Earth - 10 15 W;

All power plants in the world - 10 12 W;

All airliners - 10 11 W;

All cars - 10 5 W;

The explosion of an atomic bomb - 10 15 watts.

The conclusion suggests itself - the choice and decision-making of a person should depend, first of all, not on technical or technical

---

104

104

economic norms of society, and from socio-ethical and moral military-humanistic norms of society and the goals of its development.

Scientific and technological progress in the electric power industry is developing through the use of more and more high-calorie fuels, preproperty of forest, peat, coal, oil, gas, later nuclear topliv.

What was created by nature over millions of years of evolution is a forest massifs, biologically pure water, air, peat, oil, coal, gas - begins to partially degrade, dry out.

Here are some figures concerning the biosphere - the area, occupied by living things and vegetation. The biosphere has 2 structural parts:

- continental - $\approx$ 29% of the entire surface of the Earth, and - oceanic - $\approx$ 71% of the Earth.

Approximately 98% of the hydrosphere are seas and oceans, and only 2% - fresh water of the mainland.

By adsorbing solar energy, the plant world of the planet for a year forms $\approx$180 billion tons of biomass and $\approx$ 300 billion tons of oxygen.

Biomass serves as the basis for the nutrition of the animal world, including human, as well as the basis for the formation of wood, peat, coal, oil and gas.

The proportion of oxygen in the atmosphere decreases from fuel combustion at power plants and in boiler rooms.

An airplane, flying a thousand kilometers, uses $\approx$ 35 tons of acid kind, and during takeoff and landing, it emits so much poisonous substances, as many as 6,000 cars.

The car, running 1,000 km, spends the annual rate of oxygen consumption by humans.

More than 10 billion tons of oxygen are burned on Earth annually.

The atmosphere serves as a protective cover for the Earth, preserving it from excessive cooling and excessive heating niya.

The content of carbon dioxide CO_2 in the atmosphere should not exceed of 0.03%, but here on Earth, especially in large industrial centers, this number is exceeded by more than 3-4 times. It happens from the combustion of fossil fuels at power plants, in boiler houses, at metallurgical and other plants.

105

105

The increase in the volume of CO_2 is also due to the reduction in areas, occupied by forests, meadows; from pollution of the surfaces of the seas and oceansnew petroleum products. Reduction of land cover

goes by increasing the area for arable land, for cities and roads, artificial reservoirs.

Waste in the form of ash, dumps is deposited on the Earth's surface rocks removed from mines precipitate sulfur and other particles. Zap air density in cities is tens of times higher than in rural areas stiffness.

The seas and oceans annually receive from 6 to 12 million tons of oil and oil products due to tanker accidents, cleaning them, from oil sea wells. And one ton of oil covers up to 12 km 2 with a film water area. Already now $\approx$ 1/5 of the oil film is covered surfaces of seas and oceans.

An increase in CO 2 concentration can lead to an increase in several degrees of temperature in the low layers of the atmosphere (greenhouse effect), and this can lead to the melting of Greenland glaciers and Antarctica and the flooding of a part of the land on which 1/4 lives part of the world's population.

The world produces $\approx$ 4 billion tons of oil and gas, more than 2 billion tons of coal, about 20 billion tons of rock mass in the form of ore is extracted annually from the Earth.

The nature of the Earth, while continuing to provide humanity with the necessary dimmable resources for the development of energy production, development industry begins to change, losing its ability to be a habitat for humans.

Consider the impact of electricity production on the environment environment. Power generation is energy intensive production, and its share of impact on the environment (available in mind harmful effect) is quite significant.

4.1. TPP - thermal power plants

Fossil fuel is burned in the boiler furnaces of the TPP. Wherein oxygen is consumed and carbon dioxide - CO 2 $_{\text{is released}}$. More total CO 2 $_{\text{is}}$ emitted into the atmosphere by coal plants, a little fewer power plants running on fuel oil or natural gas. Atomic and hydroelectric power plants do not emit CO 2 .

TPPs emit into the atmosphere, in addition to CO 2 , sulphurous anhydride SO 3 , which, when combined with atmospheric moisture:

H 2 O + SO 3 = H 2 SO 4 ,

forms sulfuric acid, which falls to the Earth with rain in the form of de "acid rain". They destroy vegetation acidify the soil, cause corrosion of metal designs annoying human respiratory tract ka (Fig. 61).

At high temperatures in the boiler furnaces nitrogen oxides are formed - NO, which are also palabially affect the body

Figure: 61. Emissions into the atmosphere of industrial enterprises humans and animals. With incomplete combustion fuel injection into the atmosphere

solid particles get into, carbon monoxide gas is formed, when inhaled, in which a person can lose consciousness or even die.

Other substances are released in small quantities, harmful to humans. In addition to all these substances on coal power plants generate a fairly large amount of ash, which

Toraya takes up a lot of space, contaminates the atmosphere and contains some a certain percentage of radium isotopes, creating a certain radiation background around ash dumps.

In our country, the values of maximum permissible concentrations - MPC for various pollutants. Here are some some of them (Table 7).

Table 7

MPC of harmful emissions into the atmosphere

Substance	Chemical formula	MPC, mg / m 3
Nitrogen dioxide	NO_2	0.085
sulphur dioxide	SO_2	0.5
Carbon monoxide	CO	five
Carbon dioxide	CO_2	0.03

107
107

Excess of pollutants above the MPC is hazardous to health. IN as an instructive example, we can cite a genuine event the event that took place in London in 1952, when due to smog during 4,000 people died in 10 days (the level of SO 2 concentration on these days was in the atmosphere ≈ 4 mg / m 3).

Some power plants, such as thermal power plants and nuclear power plants, pollute reservoirs of various toxic substances - ammonia, floating howling and oxalic acids, arsenic, mercury, etc. This occurs in the chemical treatment of water for boilers and in the cleaning of boilers and pipelines during their planned repairs.

These toxic substances affect the inhabitants of the pool waters, bays, rivers and are deposited in plants and fish that humans can eat.

It is also dangerous to "thermal" contamination of the cooling water. units and systems heating up during the operation of power plants. The share of heat discharged into water bodies can be very large (up to 2/3 consumed heat of combustion of NPP fuel).

If the body of water is small, the water heats up significantly and can the inhabitants of this aquatic environment will perish, and the water will become toxic.

Nuclear power plants do not emit chemical pollutants such as thermal power plants, but they also have harmful emissions lelenia - radioactive nuclides of iodine or inert gases. They have have a short half-life, are kept for some time in gas golders, and then thrown out through the pipe of the station into the atmosphere (their activity decreases tenfold before being released).

Radioactive liquid waste from nuclear power plants is evaporated, and the dry residue
the current is poured with concrete or glass, placed in a metal container and then buried in burial grounds.

NPPs are more environmentally friendly than TPPs, but there are problems of increased danger at nuclear power plants and problems of processing and storage of spent nuclear fuel and equipment
A nuclear power plant that has served its service life.

4.2. Hydroelectric power plants

Hydroelectric power plants - hydroelectric power plants do not pollute the atmosphere with harmful secretions, do not pollute the hydrosphere and do not increase the temperature atmosphere and water bodies, but they also have a harmful effect on environment.

During the construction of a hydroelectric power station, it is necessary to provide for water reserves in the on the side of the river, i.e. it is necessary to build a reservoir. Moreover, if the terrain around the hydroelectric power station is flat, it is necessary to flood areas land, which often includes fertile land, villages and even nonbig cities.

The rise in the water level in the upper pool causes the rise of the ground return waters, waterlogging of banks, soil salinization, siltation of water ema.

The construction of a hydroelectric dam makes it difficult for fish to pass through.

usually goes to spawn against the stream of the river. Fish lifters are not very well solve this problem.

An increase in the water surface in artificial reservoirs is not rarely affects the local climate.

The construction of dams in highland areas often violates geological balance and can lead to earthquakes.

4.3. Tidal power plants, geothermal, solar stations, wind turbines, etc.

These renewable power plants energy, as if they were devoid of many of the disadvantages of thermal power plants, nuclear power plants, hydro it should be noted that due to their recent and few use has not yet come to light all the positive and negative aspects of their impact on the environment and humans. With regard to the transmission of electricity by cables and wires overhead transmission lines, they also have certain harmimpacts on humans and the environment.

Throughout the entire high-voltage transmission line is cut forest and plantations (strip width 50-100 meters), land use on this strip is excluded.

Electromagnetic radiation near power lines has harmful impact on living organisms, causes air ionization and causes interference in communication lines.

The development of energy - the main source of comfortable living of a person and the effectiveness of his life - affects

not only science, but society as a whole. Rapid population growth

The earth, the intensive development of all branches of energy, is growing impact on the environment, the finiteness of mostprimary energy resources - this is an incomplete set of problems that

109

rye need to be addressed not only for individual countries and regions, but also in worldwide.

Self-test questions

1.	What is the impact of the production of electricity and heatnew energy for the environment - Nature?
2.	What spheres surrounding the Earth can be identified? What of which are influenced by human activities?
3.	What role does the noosphere play in human activity?
4.	What are the main questions raised by V.I. Vernadsky in histhe doctrine of the biosphere and noosphere?
5.	What is the impact on the environment of the work of heatout power plants?
6.	What is the impact on the environment of the work of thestations?
7.	What is the impact on the environment of the structure andthe work of hydroelectric power stations?
8.	Do high-voltage power lines affect the environment?
editor?
9.	How can technical progress and environmental protection be combinedenvironment from harmful influences?